AF306651

André Alisson Rodrigues da Silva
Carlos A. V. de Azevedo
Geovani S. de Lima

Hydrogen peroxide in the cultivation of soursop under saline stress

André Alisson Rodrigues da Silva
Carlos A. V. de Azevedo
Geovani S. de Lima

Hydrogen peroxide in the cultivation of soursop under saline stress

Under semi-arid conditions in Northeast Brazil

ScienciaScripts

Imprint

Any brand names and product names mentioned in this book are subject to trademark, brand or patent protection and are trademarks or registered trademarks of their respective holders. The use of brand names, product names, common names, trade names, product descriptions etc. even without a particular marking in this work is in no way to be construed to mean that such names may be regarded as unrestricted in respect of trademark and brand protection legislation and could thus be used by anyone.

Cover image: www.ingimage.com

This book is a translation from the original published under ISBN 978-613-9-62989-3.

Publisher:
Sciencia Scripts
is a trademark of
Dodo Books Indian Ocean Ltd. and OmniScriptum S.R.L publishing group

120 High Road, East Finchley, London, N2 9ED, United Kingdom
Str. Armeneasca 28/1, office 1, Chisinau MD-2012, Republic of Moldova, Europe
Printed at: see last page
ISBN: 978-620-7-75900-2

SUMMARY

SUMMARY

The northeastern region of Brazil has stood out on the national scene with the production of various fruits, especially tropical fruit trees, but the concentrations of salts present in the water in these areas affect the growth and development of plants; thus making it necessary to look for alternatives for the use of these waters in irrigation; among the possibilities is the use of hydrogen peroxide, which can optimize the management of soil and/or saline water in irrigated agriculture. In this context, the aim of this study was to evaluate the emergence, growth, physiology and quality of soursop seedlings irrigated with saline water and exogenous application of hydrogen peroxide. The study was conducted in plastic bags under greenhouse conditions, using an Eutrophic Regolithic Neosol with a sandy loam texture, from the municipality of Campina Grande, PB, from May to October 2017. The treatments were distributed in a randomized block design, in a 5 x 5 factorial arrangement, with five levels of electrical conductivity of the irrigation water - ECa (0.7; 1.4; 2.1; 2.8 and 3.5 dS m^{-1}) and five concentrations of hydrogen peroxide - H2O2 (0, 25, 50, 75 and 100 µM), with four replications, the plot consisting of three plants. The concentrations of hydrogen peroxide were applied by soaking the seeds for 24 hours. Subsequently, the different concentrations of H2O2 were applied via foliar spraying on the adaxial and abaxial leaf surfaces. As salt stress increased, there was a decrease in all growth variables, with leaf area being the most affected variable. The application of hydrogen peroxide attenuated the deleterious effects of irrigation water salinity on emergence, stem diameter and leaf area. The exogenous application of hydrogen peroxide at concentrations of 25 and 50 µM attenuated the deleterious effects of water salinity on stomatal conductance, CO_2 assimilation rate and chlorophyll *a* content, with the 25 µM concentration being the most efficient. Irrigation with water from 0.7 dS m^{-1} negatively affected the quality of soursop seedlings. The concentrations of 31 and 100 µM of H O$_{22}$ promoted higher values of relative growth rate of leaf area and dry phytomass of leaf and stem, respectively. The use of saline water with an electrical conductivity of 1.22 dS m^{-1} can be used to irrigate soursop seedlings, as it promotes an acceptable average reduction of up to

10% in growth.

KEYWORDS: *Annona muricata* L., water salinity, tolerance

CHAPTER I

GENERAL INTRODUCTION AND LITERATURE REVIEW

1. GENERAL INTRODUCTION

The soursop (*Annona muricata* L.) is one of a group of fruit trees that are of economic importance to Brazil, especially the Northeast region, with the states of Bahia, Alagoas, Cearà, Paraiba and Pernambuco standing out as the largest producers. The commercial cultivation of this fruit is still recent; however, its socio-economic importance has increased in recent years due to the increased demand for tropical fruit by the food industry, as well as the possibility of its use in the pharmaceutical and cosmetics industries (Sao José et al., 2014).

Although the semi-arid region of northeastern Brazil has suitable soil and climate conditions for soursop production, it does not meet the crop's water requirements for commercial-scale cultivation due to water restrictions in terms of quantity and quality. Due to the conditions of low rainfall, irregular distribution of rainfall and intense evaporation that occurs throughout the year, high salt concentrations are common in most water sources in this region, which is one of the abiotic stress factors that ends up limiting plant growth and development (Andrade et al., 2012; Freire et al., 2014).

The use of water with high levels of salts generally causes toxicity in plant metabolism, due to the excessive accumulation of Na^+ and Cl^- ions and the reduction in the soil's osmotic potential, reducing the availability of water for the plants and consequently causing stomatal closure, limiting conductance and transpiration, which reduces the photosynthetic rate (Silva et al., 2010).

One practice that can make the use of saline water in irrigation feasible is acclimatization, carried out by pre-treating the plants with low concentrations of hydrogen peroxide. Acclimatization consists of a process in which the previous exposure of an individual to a certain type of stress causes metabolic changes that are responsible for increasing its tolerance to a new exposure to stress. Recent biochemical and genetic studies have shown that H O_{22} functions as a signaling molecule for biotic and abiotic stress, playing important roles in the developmental and physiological processes of plants, including seed germination (Barba-Espin et al., 2011), flowering (Liu et al., 2013), root system development (Hernàndez et al., 2015), and regulation of stomatal aperture (Ge et al., 2015).

Some studies have been carried out on different crops to evaluate the efficiency of hydrogen peroxide in mitigating the deleterious effects of salinity, such as rice (Uchida et al. 2002) and maize (Azevedo Neto et al. 2005), however, there is no information on pre-treatment with H2O2 and exogenous applications to soursop under conditions of saline stress.

2. OBJECTIVES

2.1. General objective

To evaluate the emergence, growth, physiology and quality of soursop seedlings irrigated with saline water and exogenous application of hydrogen peroxide, in order to provide subsidies for its cultivation in semi-arid regions.

2.2. Specific objectives

• To evaluate the emergence and growth of soursop trees irrigated with saline water and exogenous application of hydrogen peroxide concentrates;

• To analyze physiological changes by determining the gas exchange of soursop as a function of the use of saline water and hydrogen peroxide concentrations;

• To determine the levels of photosynthetic pigments in soursop seedlings under different levels of irrigation water salinity and different concentrations of hydrogen peroxide;

• To determine the concentration of hydrogen peroxide capable of mitigating the harmful effects of irrigation with water of different levels of electrical conductivity.

3. LITERATURE REVIEW

3.1. Soursop cultivation

The soursop is a crop belonging to the genus Annona, originally from the American continent, with its center of origin in Central America and the Peruvian Valleys, and is considered the most tropical of the anonaceae (Okigbo & Obire, 2009). It was spread by the Spanish and Portuguese from the Caribbean to south-eastern Mexico and Brazil (Silva et al., 2013).

It is widespread in the tropical and subtropical regions of America, Europe, Asia and Africa (Sacramento et al., 2009). It was introduced to Brazil in the 16th century, becoming one of the most economically important fruit trees, relevant in the North, Northeast, Midwest and Southeast regions, especially the states of Bahia, Alagoas, Cearà, Paraiba, Pernambuco and Parà, which have favorable soil and climatic conditions for its cultivation (Sào José et al., 2014).

Data on the area planted and the commercialization of graviola are not known exactly; however, it is known that the cultivation of this fruit has advanced in Brazil in recent years, especially in the state of Bahia, in the region of the municipality of Irecè, whose production of graviola is 8,000 tons, with prospects for growth due to the increase in planted areas in the state (ADAB, 2010).

The cultivation of soursop is of great socio-economic importance, due to the generation of employment and the retention of labor, since the production, in addition to being destined for the agro-industry, has a significant volume sold as fresh fruit, especially in the national market (Sào José et al., 2000). It is a fruit tree that has food importance and has stood out for its sensory characteristics of flavor and aroma, widely used both for *fresh* consumption and for use by the agro-industry to obtain pulp, juice, nectar, among others (Oliveira Neto et al., 2014). Furthermore, it is rich in vitamin C, calcium, carbohydrates, water and substances with antioxidant activity which have received a great deal of attention as they help protect the human body against oxidative stress, avoiding and preventing a series of chronic degenerative disorders (Taco, 2006; Yahia, 2010).

The soursop has an erect growth habit, with an average height of 4 to 8 meters in the adult phase, a single stem and asymmetrical branching. The leaves have a short petiole, are oblong-lanceolate or elliptical, 14 to 16 cm long and 5 to 7 cm wide; the veins are barely perceptible (Manica, 1997). This anonacea has the largest root system among the others and can adapt to different types of soil, although it requires deep, rich and well-drained soils with a slightly acidic pH (5.5 - 6.5) (Ramos, 1992).

The fruit varies in shape and can be ovoid, condiform or irregular, with white pulp that smells strong and sour when green, becoming soft, pleasant, juicy, cooling, sweet, slightly sour and somewhat cottony when ripe (Castro et al., 1984).

The seeds have exogenous dormancy, caused by the hardness of their outer skin, requiring scarification and/or immersion in cold water for 24 hours for perfect germination. Planting spacing can be from 4.0 x 4.5 m to 8.0 x 8.0 m (Ramos, 1992).

3.2. Salinity and its effects on crops

In general, salt stress inhibits plant growth by reducing the osmotic potential of the soil solution, limiting the availability of water and resulting in stomatal closure and, consequently, a reduction in the availability of carbon dioxide, causing damage to the photosynthetic apparatus. Furthermore, it can subject the plant to ion toxicity, nutritional imbalances or both, due to the excessive accumulation of NaCl ions in the plant tissues (Alves et al., 2011; Sà et al., 2015).

The presence of high levels of soluble salts in the rhizosphere, especially sodium, boron, bicarbonates and chlorides, causes changes in the physiological responses of plants. As a result, plants close their stomata to reduce water loss through transpiration, resulting in a lower photosynthetic rate, which is one of the causes of the reduced growth of species under conditions of salt stress (O'leary, 1975; Batista et al., 2002). On the other hand, NaCl in the soil solution interferes with the soil's physical conditions and nutrient absorption, indirectly affecting plant growth, development and metabolic activities in general (LARRÉ et al., 2014).

Tolerance to salinity is, however, variable between and even within species, depending on various factors such as phenological stage, intensity and duration of salt stress (Neves et al., 2009, Bustingorri & Lavado, 2011).

According to Cavalcante et al. (2001), the soursop tree, during the rootstock formation phase, adjusted osmotically as a moderately salt-tolerant plant, and its leaf area and the biological yield of the plants increased with the level of salinity of the irrigation water up to 3.0 dS m^{-1}.

Sà et al. (2015), when evaluating the salt balance and initial growth of pine seedlings under substrates irrigated with saline water, found that high salt concentrations in the irrigation water inhibit the emergence, growth and phytomass formation of pine plants.

Salinity levels of up to 5.5 dS m^{-1} in the irrigation water do not affect the germination of Morada Nova soursop seeds, but levels from 2.5 dS m^{-1} significantly reduce their speed of emergence (Nobre et al., 2003), which can be explained by the fact that the salts present in the irrigation water reduce the osmotic potential of the soil solution, slowing down the soaking time of the seeds.

The increase in the electrical conductivity of the soil caused by the salinity of the irrigation water negatively affects seed germination, growth and production of the yellow passion fruit, determining the need for studies that can minimize the deleterious effects of salinity and, in addition, partially or totally replace the saline supply from mineral fertilization (Cavalcante et al., 2001, 2002 and 2005).

3.2.1. Osmotic effect

When it comes to the effects caused to plants by the high concentration of soluble salts in both the soil and the irrigation water, studies must immediately turn to the osmotic component, as it is of great importance for the absorption of water by the plant (Willadino & Câmara, 2010).

Increasing the salinity of the water used for irrigation will lead to a significant increase in the salt content of the soil's saturation extract, resulting in a decrease in the osmotic potential of the soil solution, which is identified as the first factor to reduce growth (Flowers, 2004).

According to Dias & Blanco (2016), plants remove water from the soil when the soaking forces of the root tissues are greater than the forces with which water is retained in the soil. The presence of salts in the soil solution causes the retention forces to increase due to their osmotic effect, giving rise to water stress. This will have a number of implications for the plant, such as the non-absorption of nutrients, a reduction in the photosynthetic rate, a reduction in cell expansion, net CO_2 assimilation and accelerated senescence of mature leaves, consequently reducing the area allocated to the photosynthetic process and the total production of photoassimilates (Munns, 2002; Lacerda et al., 2003).

According to Munns (2002), the osmotic effect can establish a new, lower rate of leaf and root elongation in a matter of hours, causing, over time, an alteration in the start of flowering and a reduction in seed production.

Therefore, an osmotic adjustment in the plant cell is necessary to ensure the maintenance of turgor and the entry of water for cell growth. One of the mechanisms commonly cited for salinity tolerance has been the ability of some plants to accumulate ions in the vacuole and/or low molecular weight organic solutes in the cytoplasm, allowing water absorption and cell turgor to be maintained (Tester & Davenport, 2003; Taiz & Zeiger, 2017).

This compartmentalization of salt allows tolerant plants to live in saline environments, but salinity-sensitive plants tend to exclude salts in the soil solution, but are unable to carry out the osmotic adjustment described and suffer from a decrease in turgor, leading to water stress, osmosis and death (Dias & Blanco, 2016).

3.2.2. Toxic effect

When certain ions from the soil or water are absorbed by plants and accumulate in their tissues in sufficiently high concentrations, to a point where they can cause damage to the crop and reduce its yield, these plants suffer from toxicity (Silva, 2011). Generally, most crops grow under conditions of low soil salinity; therefore, the mechanisms developed to absorb, transport and utilize the nutrients present in non-saline substrates may not be effective in saline conditions (Garcia et al., 2007).

Toxicity in plants can be caused by certain elements such as sodium, boron, bicarbonates and chlorides which, in high concentrations, favor physiological disorders (Batista et al., 2002). In these toxic conditions, the concentration of Na ions$^+$ and/or Cl ions$^-$ often exceeds the concentrations of macro and micronutrients, reducing the absorption of these mineral nutrients, especially NO^3 , K^+ and Ca^{2+} (Larcher, 2000).

In saline environments, NaCl is generally the predominant salt and, consequently, the one that causes the most damage to plants. In view of this, excess Na^+ and especially Cl^- in the protoplasm cause disturbances in the ion balance, in addition to the specific effects of these ions on enzymes and cell membranes (Flores, 1990). Therefore, the deleterious effects caused by the toxicity in plants can be expressed physiologically, leading to morphological reflexes, since the high concentration of ions in the transpiratory flow causes injury to the leaves, as well as early senescence (Silva et al., 2008).

Similarly, the symptoms of toxicity by specific ions in the leaves are reported by Dias & Blanco (2016): a) the symptom of chloride is evidenced by the burning of the apex of the leaves reaching the edges in more advanced stages, promoting premature fall; b) the typical symptoms of sodium appear in the form of burns or necrosis along the edges in the older leaves, progressing in the interneval area to the center of the leaf, as it intensifies; c) the symptoms caused by boron on the leaf are summarized as yellow or dry spots on the edges and apex of the old leaves, extending through the interneural areas to the center of the leaf.

3.2.3.   Nutritional effect

In addition to the osmotic and toxic effects caused to plants by excess salts in the soil or irrigation water, it is necessary to highlight another aspect affected by salinity, which consequently affects the growth and development of crops. Nutritional imbalance occurs due to the significant alteration in the processes of absorption, transport, assimilation and distribution of nutrients in the plant; for example, excess Na^+ inhibits the absorption of nutrients such as K^+ and Ca^{++} . In addition, the high pH found in saline soil reduces the availability of many micronutrients, such as copper (Cu), iron (Fe), manganese (Mn) and zinc (Zn) (Farias et al., 2009).

When the plant is exposed to saline soil for a long time, it shows symptoms of specific ionic phytotoxicity due to the excess absorption of Na^+ or Cl^-, causing an ionic imbalance, interfering with the stomatal mechanism and causing disturbances in metabolic activities in general (Mansour & Salama, 2004).

According to BOSCO et al. (2009), salt stress caused an increase in Na^+ and Cl- ions in eggplant leaves, followed by a reduction in Ca^{2+}, Mg^{2+} and K^+, reflecting nutritional imbalance as a consequence of progressive salt stress, which also decreased the concentration of K^+ and increased the levels of N, Cu^{2+}, Na^+ and Cl^- in the stem.

3.3. Crop tolerance to salinity

Crops show different responses when subjected to a saline environment; in relation to these responses, plants can be classified into two groups: halophytes and glycophytes. Halophyte plants are those that grow naturally in environments with high saline concentrations (typically Na^+ and Cl) and can produce acceptable yields under these conditions, while glycophytes are more sensitive to relatively low salinity levels and are unable to grow in environments with high saline concentrations (Willadino & Camara, 2010).

Salt stress compromises the growth and productivity of plants all over the world and is made up of two components: osmotic and ionic (Freitas et al., 2013). The osmotic component is due to the high concentration of salts in the root environment, which causes a decrease in the soil's water potential and reduces the availability of water for the plant. The ionic component, on the other hand, is responsible for the accumulation of certain ions (generally Na^+ and cl-) and can cause nutritional imbalance, toxicity or both (Munns & Tester, 2008).

The ability of sensitive plants to survive saline stress is governed by mechanisms that confer resistance to salinity; furthermore, these mechanisms are fundamental aspects of crop growth and involve high metabolic activity under moderate stress and low activity under severe stress, which allows the plant to withstand the stress (Willidiano & Câmara, 2010).

The tolerance of plants to salinity depends on their ability to control salt transport in five specific ways: 1 - Selectivity in the absorption process by root cells; 2 - Loading the xylem preferentially with K^+, rather than Na^+; 3 - Removal of salt from the xylem in the upper part of the roots, stem, petiole or leaf sheaths; 4 - Retranslocation of Na^+ and Cl- in the phloem, guaranteeing the absence of translocation to tissues of the aerial part in the process of growth and; 5 - Excretion of salts through glands or vesicular hairs, present only in halophytes. Tolerance in glycophytes depends on the first three mechanisms, which occur to varying degrees depending on the species and/or cultivar (Munns, 2002).

One of the strategies used by plants is the extrusion of Na^+ into the soil solution, removing the cation from the plant and the expulsion of Na^+ from some tissues, especially the xylem, as a way of avoiding the accumulation of the cation in the leaf limb, minimizing the deleterious effects of salinity on leaf metabolism, especially on the photosynthetic process (Munns, 2002).There are also some species that have the capacity to accumulate ions in the vacuole and low molecular weight organic solutes in the cytoplasm, in order to lower their water potential to a level below that of the soil, which allows them to adjust osmotically to this type of condition. In turn, other crops show tolerance due to differences in the acquisition, translocation, transfer or accumulation of Na^+ and Cl^- ions (Farias et al., 2009).

Compatible solutes are a small group of chemically distinct substances, including: amino acids (such as proline), quaternary ammonium compounds (glycine betaine, β-alanine betaine, proline betaine, choline-O-sulfate), sulfonium-tertiary compounds (DMSP - dimethylsulfoniopropionate), polyols (or polyhydric alcohols such as pinitol and mannitol), soluble sugars (fructose, sucrose, trehalose, raffinose) or polymeric sugars (fructans), as well as polyamines (putrescine, spermidine and spermine). Some enzymes that eliminate free radicals and proteins that protect the formation or stability of other proteins should also be included (Willadino & Câmara, 2010).

Among the factors studied, the nutritional state of the plants is a factor that can be taken into account to characterize plant tolerance to salinity, since increases in the concentration of NaCl in the soil solution impair root absorption of nutrients, especially K^+ and Ca^+, and interfere with their physiological functions. Hence the ability of plant genotypes to maintain high levels of K and Ca and low levels of Na within the tissue is one of the key mechanisms that contribute to expressing greater tolerance to salinity. In most cases, salinity-tolerant genotypes are able to maintain high K/Na ratios in the tissues (Dias & Blanco, 2016).

3.4. Hydrogen peroxide as a salt stress attenuator

Hydrogen peroxide (H_2O_2) is one of the most stable reactive oxygen species (ROS) and is a vital component in the development, metabolism and homeostasis of different organisms (Biernert et al., 2006). However, not long ago, hydrogen peroxide and other ROS were seen only as toxic metabolites for the cell, but studies show that they act as signaling molecules, being beneficial at low concentrations and harmful when in excess (Neill et al., 2002; Uchida et al., 2002; Gechev & Hille, 2005; Quan et al., 2008).

H2O2 functions as a signaling molecule in plants under biotic and abiotic stress, playing important roles in the developmental and physiological processes of plants, including seed germination (Barba-Espin et al., 2011), flowering (Liu et al., 2013), root system development (Hernàndez et al., 2015), and regulation of stomatal opening (Ge et al., 2015).

Hydrogen peroxide can stimulate a greater accumulation of proteins and soluble carbohydrates, which will act as organic solutes, carrying out osmotic adjustment in plants under saline stress, allowing for greater water absorption (Carvalho et al. 2011). In adequate concentrations, hydrogen peroxide promotes the production of O_2 for mitochondrial respiration and metabolic activity, and can help overcome tegument dormancy, allowing better water absorption, as well as contributing to the decomposition of germination inhibitors (Oliveira Junior et al., 2017). It can also induce tolerance to salt stress by reducing the levels of Na^+ and Cl^- in plants (Gondim et al., 2011).

The pre-exposure of plants to moderate stresses or to signaling metabolites such as H_2O_2 , can result in metabolic signaling in the cell (increased metabolites and/or antioxidative enzymes) and therefore results in better physiological performance when the plant is exposed to more severe stress conditions (VEAL et al., 2007; Forman et al., 2010). In this context, several authors have observed in studies on rice (Uchida et al., 2002), maize (Azevedo Neto et al., 2005; Silva et al., 2016) wheat (Wahid et al., 2007), citrus (Tanou et al., 2009) and string beans (Hasan et al., 2016) that hydrogen peroxide acts to acclimatize plants to salt stress.

4. BIBLIOGRAPHICAL REFERENCES

ADAB. Agricultural Defense Agency of the State of Bahia. 2010 Available at: <http://www.adab.ba.gov.br/modules/news/article.php?storyid=480>. Accessed on: 21/09/2017.

Alves, M. S.; Soares, T. M.; Silva, L. T.; Fernandes, J. P.; Oliveira, M. L. A.; PAZ, V. P. S. Strategies for the use of brackish water in the production of lettuce in NFT hydroponics. Revista Brasileira de Engenharia Agricola e Ambiental, v.15, n.5, p.491 -498, 2011.

Andrade, T. S.; Montenegro, S. M. G. L.; Montenegro, A. A.; Rodrigues, D. F. B. Spatio-temporal variability of groundwater electrical conductivity in the semi-arid region of Pernambuco. Revista Brasileira de Engenharia Agricola e Ambiental, v.16, p.496-504, 2012.

Azevedo Neto, A. D.; Prisco, J. T.; Enéas Filho, J.; Rolim, M. S, J.; Gomes Filho, E. Hydrogen peroxide pre-treatment induces salt-stress acclimation in maize plants. Journal of Plant Physiology, v. 162, n. 10, p. 1114-1122, 2005.

Barba Espin, G.; Vivancos, P. D.; Job, D.; Belghazi, M.; Job, C.; Hernandez, A. J. Understanding the role of H2O2 during pea seed germination: a combined proteomic and hormone profiling approach. Plant, cell and environment, v. 34, n. 11, p. 1907-

1919, 2011.

Batista, M. J.; Novaes, F.; Santos, D. G.; Suguino, H. H. Drenagem como Instrumento de Desalinização e Prevençâo da Salinização de Solos. 2.ed., rev. e ampliada. Brasilia: CODEVASF, 216p, 2002.

Bienert, G. P.; Schjoerring, J. K.; Jahn, T. P. Membrane transport of hydrogen peroxide. Biochimica et Biophysica Acta, Amsterdam, v. 1758, n. 8, p. 994-1003, 2006.

Bosco, M. R. O. de; Oliveira, A. B. de; Hernandez, F. F. Influence of salt stress on the mineral composition of eggplant. Revista Ciência Agronòmica, v. 40, n. 2, p.157-164, 2009.

Bustingorri, C.; Lavado, R. S. Soybean growth under stable versus peak salinity. Scientia Agricola, v.68, p.102-108, 2011.

Carvalho, F. E. L.; Lobo, A. K. M.; Bonifacio, A.; Martins, M. O.; L Neto, M. C.; Silveira, J. A. G. Acclimatization to salt stress in rice plants induced by pretreatment with H_2O_2. Revista Brasileira de Engenharia Agricola e Ambiental, v.15, n.4, p.416-423, 2011.

Castro, F. A. de; Maia, G. A.; Holanda, L. F. F. Caracteristicas fìsicas e quimicas da graviola, Pesquisas Agropecuâria Brasileira, v.19, n.3, p. 361-365, 1984.

Cavalcante, L. F.; Carvalho, S. D.; Lima, E. D.; Feitosa Filho, J. C.; Silva, D. A. Initial development of soursop under sources and levels of water salinity. Revista Brasileira de Fruticultura, v.23, n.2, p.455-459, 2001.

Cavalcante, L. F.; Santos, J. B.; Santos, C. J. O.; Filho, J. C. F.; Lima, E. M. and Cavalcante, I. H. L. Seed germination and initial growth of passion fruit trees with saline water in different substrate volumes. Revista Brasileira de Fruticultura, v. 24, n. 3, p. 748-751, 2002.

Cavalcante, L. F.; Cavalcante, I. H. L.; Pereira, K S. N.; Oliveira, F. A. de; Gondim, S C.; Araùjo, F A. R. de. R. de. Germination and initial growth of guava plants irrigated with saline water. Revista Brasileira de Engenharia Agricola e Ambiental, v.9, n.4, p.515-519, 2005.

Dias, N. S.; Blanco, F. F. Effect of salts on soil and plants. In: GHEYI, H. R.; DIAS, N. S.; LACERDA, C. F. Management of salinity in agriculture: Basic and applied studies. Fortaleza, INCTA Sal, p. 151-161, 2016.

Farias, S. G. G. D.; Santos, D. R. D. U.; Freire, A. L. D. O. U. Salt stress on the initial growth and mineral nutrition of Gliricidia (Gliricidia sepium (Jacq.) Kunt ex Steud) in nutrient solution. Revista Brasileira de Ciência do Solo, v.33, n.5, p.1499-1505, 2009.

Flores, H.E.: Polyamines and plant stress. In: Alscher, R.G., Cumming, J.R. (ed.): Responses in Plants: Adaptation and Acclimation Mechanism. p. 217-239, 1990.

Flowers, T. J. Improving crop salt tolerance. Journal of Experimental Botany, v.55, p. 307-319, 2004.

Forman, H. J.; Maiorino, M.; Ursini, F. Signaling functions of reactive oxygen species. Biochemistry, v.49, p.835-842, 2010

Freitas, A. L. G. E.; Vilasboas, F. S.; Pires, M. M.; Sao José, A. R. Characterization of Graviola (Annona muricata L.) Production and Market in the State of Bahia. Informaçôes Econômicas, v. 43, n. 3, p.23-34, 2013.

Freire, J. L. O.; Dias, T. J.; Cavalcante, L. F.; Fernandes, P. D.; Lima Neto, A. J. Quantum yield and gas exchange in yellow passion fruit under water salinity, biofertilization and mulching. Revista Ciência Agronòmica, Fortaleza, v. 45, n. 1, p. 82-91, 2014.

Garcia, G. O. de; Ferreira, P. A.; Miranda, G. V. Foliar levels of cationic macronutrients and their relationship with sodium in maize plants under salt stress. IDESIA. v. 25, n.3, p. 93-106, 2007.

Ge, X. M.; Cai, H. L.; Lei, X.; Zhou, X.; Yue, M.; HE, J. M. Heterotrimeric G protein mediates ethylene-induced stomatal closure via hydrogen peroxide synthesis in Arabidopsis. The Plant Journal, v. 82, n.1, p.138-150, 2015.

Gondim, F. A.; Gomes Filho, E.; Marques, E. C.; Prisco, J. T. Effects of H2O2 on growth and solute accumulation in maize plants under salt stress. Revista Ciência Agronômica, v. 42, n. 2, p. 373-38, 2011.

Hasan, S. A.; Irfan, M.; Masrahi, Y. S.; Khalaf, M. A.; Hayat, S.; Tejada Moral, M. Growth, photosynthesis, and antioxidant responses of Vigna unguiculata L. treated with hydrogen peroxide. Cogent Food and Agriculture, v. 2, n. 1, p. 1155331, 2016.

Hernândez, A.; Velarde Buendia, A.; Zepeda, I.; Sanchez, F.; Quinto, C.; Sânchez Lopez, R.; Cardenas, L. H. Hydrogen peroxide sensor, indicates the sensitivity of the Arabidopsis root elongation zone to aluminum treatment. Sensors, v.15, n.1, p.855-867, 2015.

Lacerda, C. F.; Cambraia, J.; Cano, M. A. O.; Ruiz, H. A.; Prisco, J.T. Solute accumulation and distribution during shoot and leaf development in two sorghum genotypes under salt stress. Environmental and Experimental Botany, v.49, n.2, p.107, 2003.

Larcher, W. Ecofisiologia vegetal, Ed. RiMa Artes e Textos, Sao Carlos, 2000, p.531.

Larré, C. F.; Marini, P.; Moraes, C. L.; Amarante, L. do; Moraes, D. M. de. Influence of the24-epibrassinolide on tolerance to salt stress in rice seedlings. Semina: Ciências Agràrias, v.35, n.1, p. 67-76, 2014.

Liu, J.; Macarisin, D.; Wisniewski, M.; Sui, Y.; Droby, S.; Norelli, J.; Hershkovitz, V. Production of hydrogen peroxide and expression of ROS-generating genes in peach flower petals in response to host and non-host fungal pathogens. Plant Pathology, v.62, n.4, p. 820-828, 2013.

Manica, I. Taxonomy, morphology and anatomy. In: SÂO JOSÉ, A. R. et al. (eds.). Anonâceas, produção e mercado (pinha, graviola, atemóia and cherimólia). Vitória da Conquista: UESB. p. 20- 3, 1997.

Mansour, M.M.F.; Salama, K.H.A. Cellular basis of salinity tolerance in plants. Environmental and Experimental Botany, Elmsford, v.52, n.2, p.113-122, 2004.

Munns, R. Comparative physiology of salt and water stress. Plant, and Cell

Environment, v.25, n.2, p.239-50, 2002.

Munns, R.; Tester, M. Mechanisms of salinity tolerance. Annual Review of Plant Biology, Palo Alto, v. 59, n. 1, p. 651-681, 2008.

Neill, S. J.; Desikan, R.; Hancock, J. T. Hydrogen peroxide signaling. Current Opinion in Plant Biology, v. 5, n. 5, p. 388-395, 2002.

Neves, A. L. R.; Lacerda, C. F.; Guimaraes, F. V. A.; Hernandez, F. F.; Silva, F. B.; Prisco, J. T.; GHEYI, H. R. Biomass accumulation and nutrient extraction by string bean plants irrigated with saline water at different stages of development. Ciência Rural, v.39, n.3, p. 758-765, 2009.

Nobre, R. G.; Fernandes, P. D.; Gheyi, H. R. Santos, F. J. D. S.; Bezerra, I. L.; Gurgel, M. T. Germination and formation of grafted soursop seedlings under saline stress, Pesquisa agropecuària brasileira, v. 38, n. 12, p. 1365-1371, 2003.

Okigbo, R. N.; Obire, O. Mycoflora and production of wine from fruits of soursop (Annona Muricata L.). International Journal of Wine Research, v.1, p.1-9, 2009.

O'leary, J. W. 1975. High humidity overcomes lethal levels of salinity in hydroponically grown saltsensitive plants. Plant and Soil, v.42, n.3, p.717-721.

Oliveira Neto, E. A. de.; Santos, D. C. da.; Santos, Y. M. G. dos, Agroindustrial utilization of soursop (Annona muricata L.) for production of liqueurs: Sensory evaluation, Journal of Biotechnology and Biodiversity. v. 5, n.1, p. 33-42, 2014.

Oliveira Junior, L D de. Pre-germination treatment of forest seeds with hydrogen peroxide (Master's thesis). Federal University of Lavras, Lavras. 2017, 173p..

Quan, L. J., Zhang, B., Shi, W. W., and LI, H. Y. Hydrogen peroxide in plants: a versatile molecule of the reactive oxygen species network. Journal of Integrative Plant Biology, v.50, n.1, p.2-18, 2008.

Ramos, V. H. V. Culture of the soursop tree (Annona muricata L.) In: DONADIO, L. C. Fruticultura tropical. Jaboticabal, FUNEP, 1992, p. 268.

Sà, F. V. S.; Mesquita, E. F.; Costa, J. D.; Bertino, A. M. P.; Araùjo, J. L. Influence of gypsum and biofertilizer on the chemical attributes of a saline-sodic soil and on the initial growth of sunflower. Irriga, v. 20, n.1, p.46-59, 2015.

Sà, F. V. S. da ; Brito, M. E. B.; Pereira, I. B.; Neto, P. A.; Andrade Silva, L. de; COSTA, F. B. da. Salt balance and initial growth of pine seedlings (Annona squamosa L.) under substrates irrigated with saline water. Irriga, v. 20, n.3, p. 544, 2015.

Sacramento, C. K.; Moura, J. I. L.; Coelho Junior, E. Graviola. In: SANTOS-SEREJO, J. A. et al. (eds.). Tropical fruit growing: regional and exotic species. Brasilia, DF: Embrapa Informaçâo Tecnològica, p. 95-132, 2009.

Sao José, A. R.; Pires, M.; Freitas, A.; Ribeiro, D. P.; Perez, L. A. A. News and prospects of Anonâceas in the world. Revista Brasileira de Fruticultura, v. 36, n. especial, p. 86-93, 2014.

Sao José, A. R.; Angel, D. N.; Bonfim, M. P.; Rebouças, T. N. H. Cultivation of graviola. In: Semana Internacional de Fruticultura e Agroindùstria, v.7, 2000, Fortaleza. Courses. Fortaleza: Sindifruta, Instituto Frutal, p. 35, CD-ROM. 2000.

Silva, E. C.; Nogueira, R. J. M. C.; Araujo, F. P.; Melo N. F.; Azevedo Neto. Phisiologi- Phisiological responses to salt stress in Young umbu plants. Enviromental and Experimental botany, v. 63, p. 147-157, 2008.

Silva, I. N. Water quality in irrigation. Agricultura Cientifica no Semi Àrido, v.07, n.3, p. 01-15, 2011.

Silva, C. D. S.; Santos, P. A. A.; Lira, J. M. S.; Santana, M. C.; Silva Junior, C. D. Daily course of gas exchange in cowpea plants subjected to water deficiency. Revista Caatinga, v. 23, n. 4, p. 7-13, 2010

Silva, R. A. R. da; Nunes, J. C.; Lima Neto, A. J. de. Irrigation rates and soil cover in the production and quality of soursop fruit. Revista Brasileira Ciências Agrârias, v.8, n.3, p.441-447, 2013.

Silva, E. M. da, Lacerda, F. H. D., Medeiros, S. A. de, Souza, L. P. de, Pereira, F. H. F. Application methods of different concentrations of H_2O_2 in maize under saline stress. Green Journal of Agroecology and Sustainable Development, v.11, n. 3, p. 01-07, 2016.

Taco - Brazilian Food Composition Table. NEPA - UNICAMP - version II - 2ed - Campinas, SP: NEPA-UNICAMP, 2006.

Taiz, L.; Zeiger, E. Plant physiology. 6.ed. Porto Alegre: Artmed, 2017, 888p.

Tester, M.; Davenport, R. Na^+ tolerance and Na^+ transport in higher plants. Annals of Botany, v. 91, p. 503-527, 2003.

Tanou, G.; Job, C.; Rajjou, L.; Arc, E.; Belghazi, M.; Diamantidis, G.; Job. Proteomics reveals the overlapping roles of hydrogen peroxide and nitric oxide in the acclimation of citrus plants to salinity. The Plant Journal, v.60, n.5, p.795-804, 2009.

Uchida, A., Jagendorf, A. T., Hibino, T., Takabe, T. Effects of hydrogen peroxide and nitric oxide on both salt and heat stress tolerance in rice. Plant Science, v.*163, n.3*, p.515-523, 2002.

Veal, E. A.; Day, A. M.; Morgan, B. A. Hydrogen peroxide sensing and signaling. Molecular cell, v.26, n.1, p.1-14, 2007.

Wahid, A.; Perveen, M.; Gelani, S.; Basra, S. M. Pretreatment of seed with H2O2 improves salt tolerance of wheat seedlings by alleviation of oxidative damage and expression of stress proteins. Journal of Plant Physiology, v.164, n.3, p.283-294, 2007.

Willadino, L.; Camara, T. R. L; Plant Tolerance to Salinity: Physiological and Biochemical Aspects. Enciclopédia Biosfera, v.6, n.11; p.21, 2010.

Yahia, E. M. The contribution of fruit and vegetable consumption to human health. In L. A. Rosa, E. Alvarez-Parrilla, and G. A. Gonzalez-Aguilara (Eds.), Fruit and vegetable phytochemicals chemistry nutritional value and stability. Wiley-Blackwell: Hoboken, 2010

CHAPTER II

EXOGENOUS APPLICATION OF HYDROGEN PEROXIDE IN THE

ACCLIMATIZATION OF SOURSOP SEEDLINGS TO SALT STRESS

ABSTRACT: The northeastern region of Brazil has favorable conditions for the exploitation of various crops, but the high concentration of salts in irrigation water is often a limiting factor for production. The aim of this study was to evaluate the emergence, growth and partitioning of photoassimilates of Morada Nova soursop seedlings irrigated with saline water and exogenous application of hydrogen peroxide through seed soaking and foliar spraying. The study was conducted in plastic bags under greenhouse conditions, using an Eutrophic Regolithic Neosol with a sandy loam texture, from the municipality of Campina Grande, PB. The treatments were distributed in a randomized block design, in a 5 x 5 factorial arrangement, with five levels of electrical conductivity of the irrigation water - ECa (0.7; 1.4; 2.1; 2.8 and 3.5 dS m^{-1}) associated with five concentrations of hydrogen peroxide - H O$_{22}$ (0, 25, 50, 75 and 100 µM), with four replications and three plants per plot. As salt stress increased, there was a decrease in all the variables analyzed, with leaf area being the most affected. The application of hydrogen peroxide attenuated the deleterious effects of irrigation water salinity on the percentage of emergence, emergence speed index, stem diameter and leaf area, with the concentration of 50 µM being the most efficient.

Keywords: *Annona muricata* L., saline water, osmoregulator

1. INTRODUCTION

Fruit growing is an activity of great importance to the Brazilian agricultural sector. In recent years, Brazil has excelled on a global level, being among the world's top three fruit producers, behind only China and India, respectively. According to the Brazilian Fruit Yearbook, the estimated fruit production in 2017 was approximately 44 million tons; it is estimated that the fruit production chain covers three million hectares and generates six million direct jobs, thus highlighting the importance of fruit growing in Brazilian agribusiness.

The soursop tree (*Annona muricata* L.), which belongs to the anonaceae family, has great economic value, especially in the Northeast region of Brazil, where it is grown on a large scale due to the favorable soil and climatic conditions for its cultivation, as well as being appreciated by the population for its flavor, aroma and pharmaceutical characteristics (Freitas et al., 2013).

Although the Northeast region of Brazil has favorable soil and climate conditions for the production of soursop, this is not enough to exploit the potential of this crop, due to the limitations imposed by the rainfall regimes of this region (Silva et al., 2015), 2015), characterized by prolonged periods of drought and irregular annual rainfall, causing a water deficit for the plants, since the rate of evapotranspiration exceeds that of precipitation for most of the year (Holanda et al., 2016), a situation which favours the increase in saline levels in the water sources available for irrigation.

Salt stress inhibits plant growth due to a reduction in the osmotic potential of the soil solution, limiting the availability of water and resulting in stomatal closure and, consequently, a reduction in the availability of carbon dioxide, causing damage to the photosynthetic apparatus. In this way, the plant can be subjected to ion toxicity, nutritional imbalances or both, due to the excessive accumulation of Na ions$^+$ and Cl_- in plant tissues (Alves et al., 2011; Sà et al., 2015).

It should be considered that the formation of soursop seedlings in the semi-arid region of northeastern Brazil can be optimized with the use of techniques that make soil and/or saline water management feasible in agriculture. Among these alternatives, the exogenous application of hydrogen peroxide (H_2O_2) has shown promise in mitigating the effects of salt stress on crops (Oliveira, 2016; Gondim et al., 2011; Carvalho et al., 2011).

In this context, several authors have observed in studies carried out with rice (Uchida et al., 2002), maize (Azevedo Neto et al., 2005; Silva et al., 2016) wheat (Wahid et al., 2007), orange (Tanou et al., 2009) and string bean (Hasan et al., 2016), 2016) that hydrogen peroxide acts to acclimatize plants to salt stress; however, for the soursop crop there is no information on the use of hydrogen peroxide to attenuate or acclimatize the crop to salt stress conditions. Hydrogen peroxide can stimulate a greater accumulation of proteins and soluble carbohydrates, which will act as organic solutes, carrying out the osmotic adjustment of plants under salt stress, allowing for greater water absorption (Carvalho et al., 2011). The aim of this study was to evaluate the emergence, growth and partitioning of photoassimilates of Morada Nova soursop seedlings irrigated with saline water and exogenous application of hydrogen peroxide.

2. MATERIAL AND METHODS

The work was carried out from May to October 2017, in plastic bags measuring 2 dm^3 , under greenhouse conditions, belonging to the Center for Technology and Natural Resources of the Federal University of Campina Grande (CTRN/UFCG), located in Campina Grande, PB, at geographic coordinates 07° 15' 18" S latitude, 35° 52' 28" W longitude and an average altitude of 550 m.

The treatments resulted from the combination of two factors: five levels of electrical conductivity of the irrigation water - ECa (0.7; 1.4; 2.1; 2.8 and 3.5 dS m^{-1}) associated with five concentrations of hydrogen peroxide - H2O2 (0, 25, 50, 75 and 100 µM), distributed in a randomized block design, in a 5 x 5 factorial arrangement, with four replications and three plants per plot, making a total of three hundred experimental units.

The levels of electrical conductivity of the irrigation water (1.4; 2.1; 2.8 and 3.5 dS m^{-1}) were prepared by dissolving the salts NaCl, CaCl2.2H2O and MgCl2.6H2O, in the equivalent ratio of 7:2:1, between Na^+ , Ca^2 + and Mg^2 +, respectively, in local supply water (ECa = 1.10 dS m^{-1}). This proportion of salts is commonly found in water sources used for irrigation in small properties in the Northeast of Brazil (Medeiros et al., 2003), based on the relationship between ECa and the concentration of salts (10*mmolc L^{-1} = ECa dS m^{-1}), taken from Richards (1954). The level of 0.7

dS m^{-1} was obtained by diluting the local water supply with rainwater (ECa = 0.02 dS m).$^{-1}$

The plastic bags were filled with 2.6 kg of a substrate made up of soil (84%) + sand (15%) + humus (1%). The soil used in the experiment was Neossolo Regolitico Eutròfico with a sandy loam texture, collected at a depth of 0-20 cm from the rural area of the municipality of Lagoa Seca, PB, and duly crushed and sieved. Its physical, hydric and chemical characteristics (Table 1) were determined according to the methodology proposed by Donagema et al. (2011).

Table 1. Chemical and physico-hydric attributes of the soil used in the experiment, before the treatments were applied.

Chemical characteristics									
pH (H2O) (1:2, 5)	M.O. %	P (mg kg)$^{-1}$	K$^+$	In$^+$	Ca^{2+} .... (cmolckg)$^{-1}$	Mg $+^2$	Al^{3+} + H$^+$	PST (%)	CEes (dS m^{-1})
5,90	1,36	6,80	2,22	1,60	26,00	36,60	19,30	1,87	1,0

Physico-hydric characteristics								
Particle size (dag)			Class textural	Humidity (kPa) 33,42	1519.5 dag kg^{-1}	AD	Total porosity %	DA (g cm)$^{-3}$	DP
Sand	Silt	Clay						
73,29	14,21	12,50	FA	11,98	4,32	7,66	47,74	1,39	2,66

O.M. - Organic matter:Walkley-Black wet digestion; Ca^{2+} and Mg^{2+} extracted with KCl 1 mol L^{-1} pH 7.0; Na$^+$ and K$^+$ extracted using NH4OAc 1 mol L^{-1} pH 7.0; Al^{3+} and H$^+$ extracted with calcium acetate 1 mol L^{-1} pH 7.0; PST- Percentage of exchangeable sodium; ECes - Electrical conductivity of the saturation extract; FA - Sandy loam; AD - Available water; DA- Apparent density; DP- Particle density.

The seeds used in the experiment were obtained from fruit harvested from a commercial orchard located in the municipality of Macaparana, PE. The seeds were extracted manually, then air-dried and the process of breaking dormancy was carried out by cutting them distal to the embryo, according to the methodology proposed by Mendonça et al. (2007).

The different concentrations of hydrogen peroxide (H O$_{22}$), previously established according to the studies, were obtained by diluting H O$_{22}$ in deionized water. Before sowing, the seeds underwent a pre-treatment with hydrogen peroxide, where they were soaked in the concentrations of the respective treatments for a period of 24 hours; the seeds were then sown by placing three seeds at a depth of 3 cm and distributed equidistantly; at 20 days after germination, thinning was carried out in order to have only one plant per bag, leaving the one with the best vigor.

Before sowing, the soil moisture content was raised to field capacity using the water for each treatment. After sowing, irrigation was carried out daily by applying a volume of water to each plastic bag in order to maintain soil humidity close to field capacity. The volume applied was determined according to the plants' water requirements, estimated by the water balance by subtracting the volume drained from the volume applied during the previous irrigation, plus a leaching fraction of 0.10. Top dressing was applied with nitrogen, potassium and phosphorus, based on the methodology contained in Novais et al. (1991). 0.58 g of urea, 0.65 g of potassium chloride and 1.56 g of monoammonium phosphate were applied, equivalent to 100, 150 and 300 mg kg^{-1} of the substrate of N, K and P, respectively, applied as top dressing in four applications via fertigation, at 15-day intervals, with the first application made 15 days after sowing (DAS); to make up for possible micronutrient deficiencies, 2.5 g L^{-1} of ubyfol [(N (15%); P2O5 (15%); K2O (15%); Ca (1%); Mg (1.4%); S (2.7%); Zn (0.5%); B (0.05%); Fe (0.5%); Mn (0.05%); Cu (0.5%); Mo (0.02%)] was applied foliarly at 60 and 100 DAS. At 90, 105 and 120 DAS, foliar spraying with the appropriate hydrogen peroxide solutions was carried out by hand using a spray bottle.

The effects of the different ECa levels and hydrogen peroxide concentrations on soursop seedlings cv. Morada Nova at 85, 100 and 145 DAS were determined using the percentage of seedling emergence (PE), emergence speed index (IVE), plant height (AP), stem diameter (DC), number of leaves (NF), leaf area (AF), net assimilation rate (TAL), specific leaf area (AFE), leaf area ratio (RAF) and leaf succulence (SUC).

The percentage of seedling emergence was obtained by counting the number of emerged seedlings every day until they were established, using the criterion of the emergence of the epicotyl on the surface of the container. Once this data was available, the IVE (seedlings day^{-1}) was determined using Equation 1 presented by

Carvalho and Nakagawa (2000):

$$IVE \left(pl\hat{a}ntulas\ dia^{-1}\right) = \frac{\sum_1}{N_1} + \frac{\sum_2}{N_2} \ldots + \frac{\sum_n}{\sum_n} \qquad (1)$$

Where:

$\sum_1$, $\sum_2$, ... $\sum_n$ number of seedlings emerged, respectively, in the first, second, ... and last counts; and,

N_1 , N_2 , ... N_n - number of days between sowing and the first, second, ... and last count, respectively.

The variable plant height (cm) was measured by taking as a reference the distance from the neck of the plant to the insertion of the apical meristem, the DC (mm) was measured at 2 cm from the neck of the plant and the number of leaves was obtained by counting the fully expanded leaves with a minimum length of 3 cm on each plant.

The leaf area (cm^2) was determined according to the recommendations of Almeida et al. (2006), using Equation 2:

$$AF = 5,71 + 0,647X \qquad (2)$$

Where:

AF - leaf area (cm^2); e,

X - product of the length and width of the leaves (cm).

The net assimilation rate (NAR) (g cm^{-2} day^{-1}) was determined in the periods between two evaluations (85 and 145 DAS). The specific leaf area (SFA) (cm^2 g^{-1}) and the leaf area ratio (LAR) (cm^2 g^{-1}) were measured at 145 DAS, according to the methodology proposed by Benincasa (2003):

$$TAL = (MS_2 - MS_1)x(LnAF_2 - LnAF_1)\ x\ (t_2 - t_1)^{-1}\ x\ (AF_2 - AF_1)^{-1} \qquad (3)$$

Where:

DM_1 - total dry mass at time 1 (g);

DM_2 - total dry mass at time 2 (g);

AF_1 - total leaf area at time 1 (cm^2);

AF_2 - total leaf area at time 2 (cm^2);

t_1 - collection at time 1 (days); e,

t_2 - collection at time 2 (days):

$$AFE = \frac{AF}{MSF} \qquad (4)$$

$$RAF = \frac{AF}{MST} \qquad (5)$$

Where:

AF - total leaf area (cm^2);

MSF - leaf dry mass (g); e,

MST - total dry mass (g).

Leaf succulence (SUC) was determined at 145 DAS, according to the methodology proposed by Mantovani (1999), given by Equation 6:

$$SUC = \frac{(FF - FS)}{AF} \qquad (6)$$

Where:

FF - fresh phytomass (g);

FS - dry phytomass (g); e,

AF - leaf area (cm^2)

The data collected was subjected to analysis of variance using the F test at 0.05 and 0.01 probability levels and, when significant, regression analysis linear and quadratic polynomial, using the SISVAR statistical software (Ferreira, 2014).

3. RESULTS AND DISCUSSION

The germination process of soursop seeds, as assessed by the speed of seedling emergence (IVE) and the percentage of emerged seedlings (PE), was significantly affected (p < 0.01) by the interaction between the factors studied (Table 2). When graviola cv. Morada Nova plants were irrigated with saline water, their IVE and PE were significantly affected (p < 0.01). The concentrations of hydrogen peroxide (H_2O_2) had a significant effect (p < 0.01) only on IVE. Killic & Kahraman (2016), when studying the effect of exogenous application of hydrogen peroxide on barley seedlings subjected to salt stress, also observed a significant effect of the interaction between these factors.

Table 2. Summary of the F-test for percentage emergence (PE) and emergence speed index (IVE) of Morada Nova soursop irrigated with saline water and exogenously applied concentrations of hydrogen peroxide.

	F-test	
Variation source	PE	IVE
Salt Levels (NS)	**	**
Linear regression	ns	ns
Quadratic regression	ns	*
Hydrogen peroxide (H_2O_2)	ns	**
Linear regression	*	**
Quadratic regression	ns	ns
Interaction (NS x H_2O_2)	**	**
Blocks	ns	ns
CV (%)	9,82	12,94

[**] ns, **, * respectively not significant, significant at p < 0.01 and p < 0.05.

The regression equation (Figure 1A) shows a reduction in IVE with an increase in the electrical conductivity of the irrigation water in plants that were not treated with hydrogen peroxide (control), corresponding to a decrease of 5.46% per unit increase in salinity. However, when the concentrations of hydrogen peroxide were applied, it was clear that the deleterious effect on the germination process caused by water salinity was mitigated, and the regression equation (Figure 1A) showed that the concentration of H_2O_2 that resulted in the highest IVE (0.076 seedlings day^{-1}) was 50 µM, even when associated with the highest water salinity level of 3.5 dS m^{-1} .

For the percentage of emergence (PE) of soursop seedlings, a decreasing linear behavior was observed as a function of the salinity levels of the irrigation water in the control seedlings (0 µM); according to the regression study (Figure 1B), there was a 17.7% reduction in PE in seedlings irrigated with the water with the highest salinity

(3.5 dS m^{-1}) compared to those with the lowest salinity level (0.7 dS m^{-1}). Seeds pre-treated with concentrations of hydrogen peroxide and exposed to saline levels in the irrigation water had higher PE, indicating that adequate levels of H2O2 promote acclimatization to saline stress, and the regression equations (Figure 1B) show that even using water with an ECa of 3.5 dS m^{-1} associated with a concentration of 75 μM of H2O2 resulted in a higher percentage of emergence (91.4%).

The positive effect of applying hydrogen peroxide on the IVE and PE of soursop can be attributed to the fact that these concentrations of H O$_{22}$ in the cells cause o2 production for mitochondrial respiration and metabolic activity. According to Oliveira Junior (2017), hydrogen peroxide can help overcome tegument dormancy, allowing for better water absorption, as well as contributing to the decomposition of germination inhibitors.

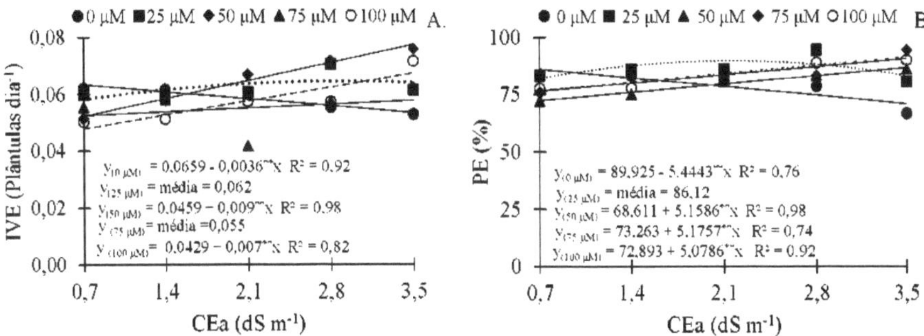

Figura 1.   emergence speed index - IVE (A) and emergence percentage - PE (B) of soursop cv. Morada Nova as a function of the interaction between irrigation water salinity and hydrogen peroxide concentrations.

The reductions in IVE and PE with the increase in salinity in the control plants (level 0 of H O$_{22}$) can be attributed to the reduction in osmotic potential caused by the concentration of soluble salts in the soil; in this way there is less water absorption by the plants, as well as the entry of ions in sufficient quantities to cause toxicity to the embryo and/or endosperm membrane cells; In general, toxic concentrations of these ions (Na^{+} and Cl^{-}) affect other processes, including cell division and differentiation, enzymatic activities and nutrient distribution, which can delay seedling emergence and the mobilization of reserves, contributing to a reduction in seed viability (Lima et al., 2014, Sà et al., 2015).

The F test (Table 3) showed significant effects of the salinity levels of the irrigation water on plant height (PH), stem diameter (SD), number of leaves (NF) and leaf area at 85, 100 and 145 DAS. In relation to the concentrations of hydrogen peroxide, there was a significant effect on PA and NF at 85 and 145 DAS, on DC at 85 and 100 DAS, and on AF at 145 DAS; moreover, the interaction between salt levels and H2O2 had a significant effect on DC at 85 DAS and on AF at 145 DAS.

Table 3. Summary of the F test for plant height (PH), stem diameter (SD), number of leaves (NF) and leaf area (LA) of Morada Nova soursop plants irrigated with saline water and exogenous hydrogen peroxide at 85, 100 and 145 days after sowing (DAS).

	F-test											
	AP			DC			NF			AF		
	Days after sowing											
Variation source	85	100	145	85	100	145	85	100	145	85	100	145
Salt Levels (NS)	**	**	**	**	**	**	**	**	**	**	**	**
Linear regression	**	**	**	**	**	**	**	**	**	**	**	**
Quadratic regression	ns	ns	ns	ns	ns	ns	ns	ns	ns	ns	ns	ns
Hydrogen Peroxide (H_2O_2)	**	ns	**	**	**	ns	**	ns	**	ns	ns	**
Linear regression	**	ns	**	**	**	ns	**	ns	**	ns	ns	ns
Quadratic regression	ns	ns	ns	ns	ns	ns	ns	ns	ns	ns	ns	**
Interaction (NS x H_2O_2)	ns	ns	ns	**	ns	ns	ns	ns	ns	ns	ns	**
Blocks	ns	ns	ns	ns	ns	ns	ns	ns	ns	ns	ns	ns
CV (%)	7,78	8,57	7,14	6,25	5,06	8,31	20,90	10,28	5,72	16,41	14,34	8,42

ns, ** respectively not significant and significant at $p < 0.01$.

The increase in ECa negatively affected the PA (Figure 2A) and DC (Figure 2D) of the Morada Nova soursop at 85, 100 and 145 DAS. According to the regression equations, there were declines in PA of 4.28, 4.90 and 6.31% per unit increase in ECA at 85, 100 and 145 DAS, respectively, and in DC of 3.60 and 6.03% per unit increase in the evaluation carried out at 100 and 145 DAS, respectively. The reductions in PA and DC as a result of salinity may be related to water deficiency, induced by the osmotic effect, promoting the closure of stomata and a reduction in gas exchange, consequently reducing the absorption of water and nutrients by the plants, which results in lower growth (Lima et al., 2015).

The different concentrations of hydrogen peroxide had a positive effect on the PA at

85 and 145 DAS (Figure 2B) and on the DC at 100 DAS (Figure 2E). The regression equations showed an increasing linear effect, with an increase in the PA of 8.51 and 7.87% at 85 and 145 DAS respectively, and an increase in the DC of 7.38% compared to the plants that did not receive the H2O2 treatment.

This response demonstrates the efficiency of exogenous application of hydrogen peroxide (H2O2) in acclimatizing plants to salt stress. H2O2 functions as a signaling molecule in plants under conditions of biotic and abiotic stress (Petrov & Breusegem, 2012), and when applied in low concentrations in plants it induces the antioxidative enzyme defense system, minimizing the deleterious effects of salinity (Carvalho et al., 2011). In addition, it can induce tolerance, promoting the accumulation of soluble proteins, soluble carbohydrates and NO_3^- as well as reducing the levels of Na^+ and Cl^- in plants (Gondim et al., 2011).

It can be seen that the diameter of the plants that did not receive exogenous application of H2O2 was reduced as the electrical conductivity of the irrigation water increased (Figure 2C), with a linear decrease of 6.26% per unit increase in ECa, i.e. an 18.34% reduction in DC in plants irrigated with the water with the highest salinity (3.5 dS m^{-1}) in relation to the lowest salinity level (0.7 dS m^{-1}). However, the negative effect caused by salinity was mitigated in the Morada Nova soursop plants when they were treated with hydrogen peroxide at a concentration of 50 µM associated with an electrical conductivity of 2.1 dS m^{-1} , obtaining a diameter of 2.77 mm, representing an increase of 7.07% compared to the control treatment (0 µM).

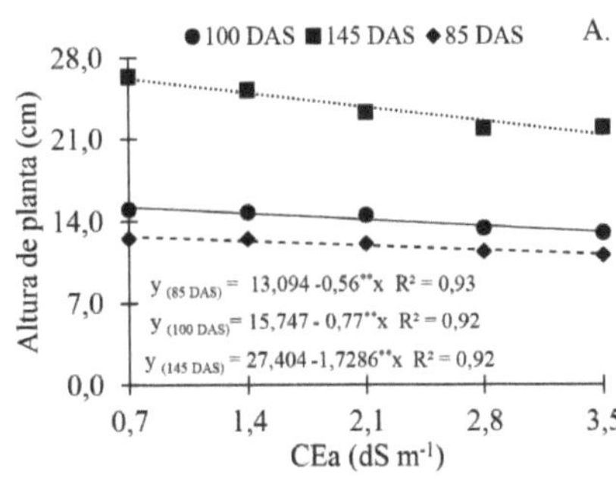

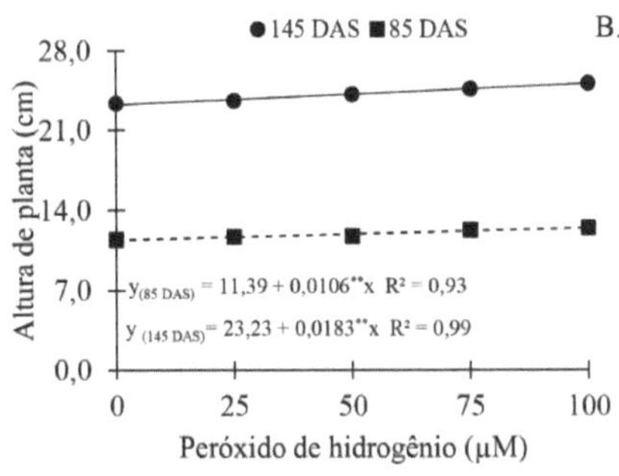

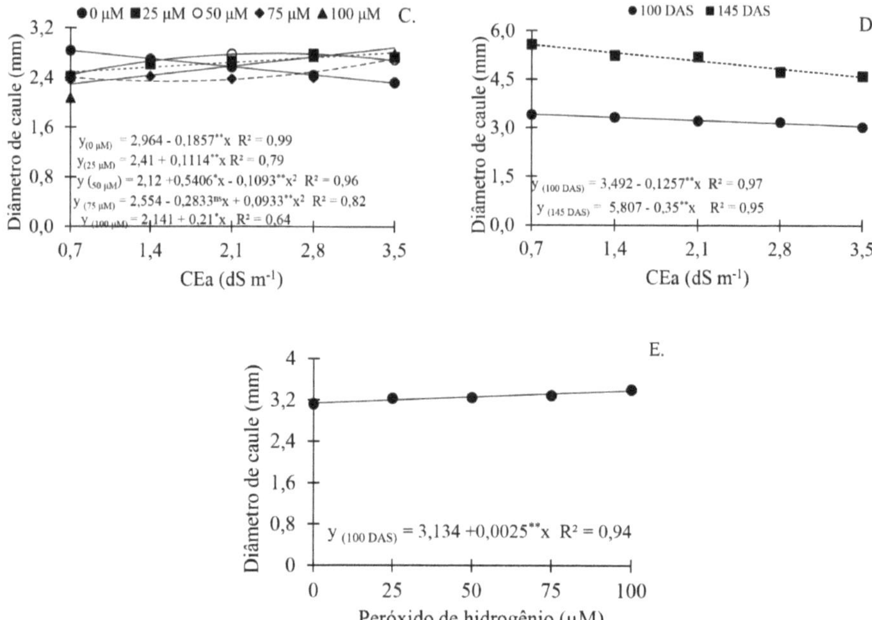

Figura 2. Plant height of soursop cv. Morada Nova at 85, 100 and 145 days after sowing (DAS) as a function of salinity (A) and at 85 and 145 DAS as a function of H O$_{22}$ (B); stem diameter at 85 DAS as a function of the salinity x H2O2 interaction (C) and at 100 and 145 DAS as a function of irrigation water salinity (D), and at 100 DAS as a function of hydrogen peroxide concentrations (E).

As for the number of leaves (NF) of the soursop, irrigation with increasingly saline water led to reductions in this variable; based on the regression studies (Figure 3A), there was a decreasing linear effect on NF, with decreases in the order of 6.66%, 5.25% and 7.10% per unit increase in ECa at 85, 100 and 145 DAS, respectively. This result may be a consequence of the plant's adaptation mechanisms to salt stress, reducing the transpiring surface. In this way, the reduction in the number of leaves under such conditions is relevant to maintaining a high water potential in the plant (Nobre et al., 2014). Concentrations of H2O2 had a positive effect on NF at 85 and 145 DAS (Figure 3B), with increases of 17.67% and 10.45% at the highest level of salinity, respectively, at 85 and 145 DAS, compared to plants submitted to 0 μM (control).

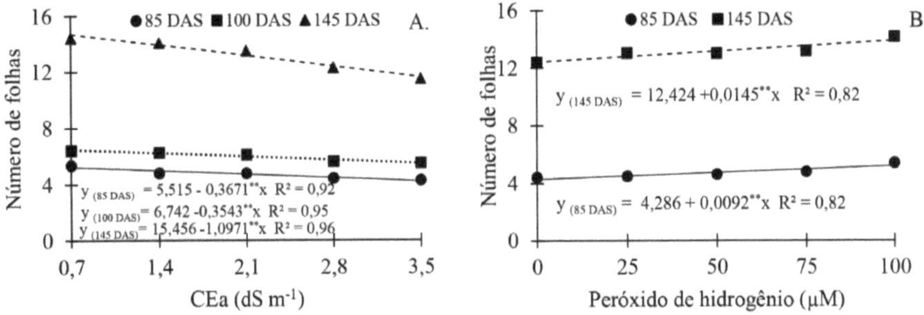

Figura 3. Number of leaves of Morada Nova soursop at 85, 100 and 145 days after sowing (DAS) as a function of salinity (A) and at 85 and 145 DAS as a function of hydrogen peroxide concentrations (B).

Analyzing the regression equation (Figure 4A) for leaf area (LA) at 85 and 100 DAS, the linear model indicates decreases of 7.79 and 6.61% per unit increase in ECa, respectively, i.e. a reduction of 23.07% (22.58 cm^2) at 85 DAS and 19.40% (24.43 cm^2) at 100 DAS in plants irrigated with the highest salinity water (3.5 dS m^{-1}) compared to the lowest saline level (0.7 dS m)$.^{-1}$

The concentrations of H O$_{22}$ of 25 and 50 μM had a beneficial effect on AF at 145 DAS, showing that the deleterious effects caused by water salinity were mitigated, and according to the regression studies (Figure 4B) it can be seen that the highest AF (311.82 cm^2) was obtained at a concentration of 50 μM associated with salinity of 2.1 dS m^{-1} . However, with regard to the hydrogen peroxide concentrations of 75 and 100 μM, it can be seen that they intensified the negative effect caused by the increase in the electrical conductivity of the irrigation water, possibly because they are toxic to the Morada Nova soursop tree. High concentrations of this reactive oxygen species induce oxidative stress, causing lipid peroxidation, damage to cell membranes, protein degradation, DNA double-strand breaks and cell death (Miller et al., 2007; Nguyen et al., 2009; Rutschow et al., 2011).

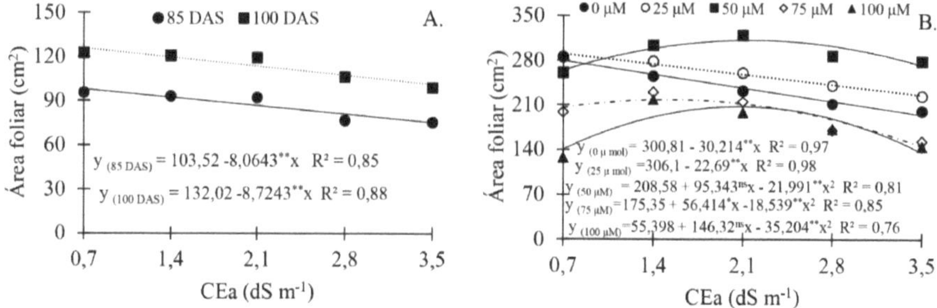

Figure 4 - Leaf area of soursop cv. Morada Nova at 85 and 100 days after sowing (DAS) as a function of salinity (A) and at 145 DAS as a function of the interaction between electrical conductivity of irrigation water and hydrogen peroxide concentrations.

The summary of the F test (Table 4) shows that the saline levels of the irrigation water significantly influenced ($p < 0.01$) only the leaf succulence variable (SUC). The concentrations of hydrogen peroxide significantly affected the specific leaf area (SFA) and SUC ($p < 0.01$) as well as the net assimilation rate (NAR) ($p < 0.05$). However, there was no significant effect of the interaction (salt levels x H_2O_2) for any of the variables analyzed ($p > 0.05$).

Table 4 Summary of the F-test for specific leaf area (SFA), leaf area ratio (LAR), net assimilation rate (NAR) and leaf succulence (LAS) of Morada Nova soursop irrigated with saline water and exogenous application of hydrogen peroxide.

Source of Variation	F-test			
	AFE	RAF	SUCH	SUC
Salt Levels (NS)	ns	ns	ns	**
Linear regression	ns	ns	**	**
Quadratic regression	ns	ns	ns	ns
Hydrogen Peroxide (H_2O_2)	**	ns	*	**
Linear regression	ns	ns	ns	**
Quadratic regression	**	ns	**	ns
Interaction (NS x H_2O_2)	ns	ns	ns	ns
Blocks	ns	ns	ns	ns
CV (%)	15,6	18,22	21,33	24,03

ns, **, * respectively not significant, significant at $p < 0.01$ and $p < 0.05$.

The concentrations of hydrogen peroxide influenced the AFE variable at 145 DAS, with the data fitting the quadratic model (Figure 5A), with the maximum value estimated at 239.86 cm^2 g^{-1} in the plants submitted to 50 μM of hydrogen peroxide, with depletion occurring from this concentration onwards. The minimum value found was 169.68 cm^2 g^{-1} in plants subjected to a concentration of 100 μM; this response shows that the application of a high concentration of hydrogen peroxide causes damage to plants, possibly due to the alterations that occur in plant metabolism, above all as a consequence of oxidative stress, causing a restriction in photosynthetic processes (Cattivelli et al., 2008).

The different concentrations of H2O2 influenced the net assimilation rate (NAR) of the soursop cv. Morada Nova, and according to the regression equation (Figure 5B) it can be seen that the plants submitted to treatment with hydrogen peroxide at a concentration of 50 μM stood out with the highest TAL value (0.000138 g cm^2 day^{-1}) in relation to the other treatments, following the same trend observed for the specific leaf area.

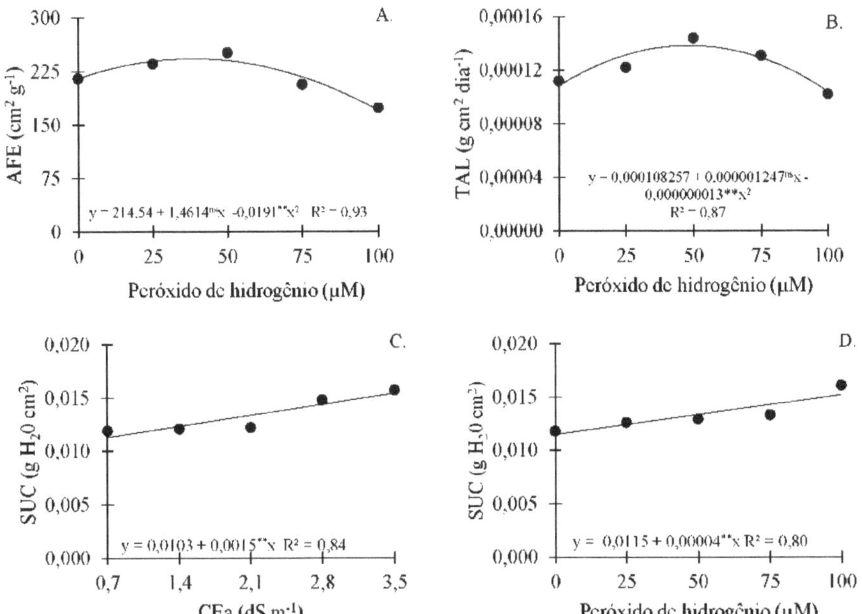

Figure 5 - Specific leaf area - SFA of soursop cv. Morada Nova (A) and net assimilation rate - NAR (B) as a function of salinity, and leaf succulence - SUC as a

function of salinity (C) and hydrogen peroxide concentrations (D).

With regard to the leaf succulence (SUC) of the soursop, there was a linear increase in response to the increase in irrigation water salinity. According to the regression equation (Figure 4C), there was a 14.56% increase in SUC per unit increase in ECa, equivalent to an increase of 37.0% (0.0042 g H_2 O cm^2), in plants irrigated with ECa of 3.5 dS m^{-1} , compared to those subjected to water salinity of 0.7 dS m^{-1} . Thus, the increase in leaf succulence caused by salt stress in soursop may be an indication that osmotic adjustment has occurred in the plants. Silva et al. (2009) also found that there was an increase in the leaf succulence of jatropha plants grown under salt stress (0, 25, 50, 75 and 100 mmolc L^{-1} of NaCl); the authors also mentioned that this increase in SUC played an effective role in the osmotic adjustment of the plants.

The different concentrations of hydrogen peroxide also influenced SUC, with a linear increase in response to the increase in hydrogen peroxide concentration. The regression equation (Figure 4D) shows an increase of 0.35% per increase in hydrogen peroxide concentration.

4. CONCLUSIONS

Increased water salinity reduces the percentage of emergence, the emergence rate, growth and partitioning of photoassimilates in Morada nova soursop.

The exogenous application of hydrogen peroxide to soursop attenuates the deleterious effects of irrigation water salinity on emergence, stem diameter at 85 days after sowing and leaf area at 145 days after sowing, with a concentration of 50 µM being the most efficient.

Concentrations of hydrogen peroxide above 50 µM reduce the specific leaf area of the Morada nova soursop tree.

5. BIBLIOGRAPHICAL REFERENCES

Brazilian Fruit Yearbook. Brazilian fruits yearbook. Santa Cruz do Sul, Gazeta, p. 88p, 2017.

Almeida, G.; Santos, J.; Zucoloto, M.; Vicentini, V.; Moraes, W., Bregoncio, I.;

Coelho, R. Estimation of leaf area of graviola (*Annona muricata* L.) by means of linear dimensions of the leaf limb. Revista UNIVAP, v. 1, p.1035-1037, 2006.

Alves, M. S.; Soares, T. M.; Silva, L. T.; Fernandes, J. P.; Oliveira, M. L. A.; Paz, V. P. S. Strategies for using brackish water in the production of lettuce in NFT hydroponics.

Revista Brasileira de Engenharia Agricola e Ambiental, v.15, p.491-498, 2011.

Arnon, D. I. Copper enzymes in isolated chloroplasts: polyphenoloxidases in *Beta vulgaris*. Plant Physiology, v.24, p.1-15, 1949.

Azevedo Neto, A. D.; Prisco, J. T.; Enéas Filho, J.; Rolim, M. S, J.; Gomes Filho, E. Hydrogen peroxide pre-treatment induces salt-stress acclimation in maize plants. Journal of Plant Physiology, v. 162, p. 1114-1122, 2005.

Benincasa, M. M. P. Plant growth analysis, basics. 2.ed.

Jaboticabal: FUNEP, 2003. 41p.

Carvalho, N. M.; Nakagawa, J. Seeds: Science, technology and production. 4.ed. Jaboticabal: FUNEP, 2000, 588p.

Carvalho, F. E. L.; Lobo, A. K. M.; Bonifacio, A.; Martins, M. O.; L Neto, M. C.; Silveira, J. A. G. Acclimatization to salt stress in rice plants induced by pretreatment with H_2O_2. Revista Brasileira de Engenharia Agricola e Ambiental, v.15, p.416-423, 2011.

Cattivelli, L.; Rizza, F.; Badeck, F. W.; Mazzucotelli, E.; Mastrangelo, A. M.; Francia, E.; Maré, C.; Tondelli, A.; Stanca, A. M. Drought tolerance improvement in crop plants: An integrated view from breeding to genomics. Field Crops Research, v.105, p.1-14, 2008

Donagema, G. K.; Campos, D. V. B. de; CALDERANO, S. B.; TEIXEIRA, W. G.; VIANA, J. H. M. (Org.). Manual of soil analysis methods. 2.ed. Rio de Janeiro: Embrapa Solos, 2011.

Ferreira, D. F. Sisvar: A guide for its bootstrap procedures in multiple comparisons.

Ciência e Agrotecnologia, v. 38, p. 109-112, 2014.

Freitas, A. L. G. E.; Vilasboas, F. S.; Pires, M. M.; Sào José, A. R. Characterization of graviola (*Annona muricata* L.) production and market in the State of Bahia. Informaçôes Econômicas, v. 43, p.23-34, 2013.

Gondim, F. A.; Gomes Filho, E.; Marques, E. C.; Prisco, J. T. Effects of H2O2 on growth and solute accumulation in maize plants under salt stress. Revista Ciência Agronômica, v. 42, p. 373-38, 2011.

Hasan, S. A.; Irfan, M.; Masrahi, Y. S.; Khalaf, M. A.; Hayat, S.; Tejada Moral, M. Growth, photosynthesis, and antioxidant responses of *Vigna unguiculata* L. treated with hydrogen peroxide. Cogent Food and Agriculture, v. 2, p. 1155331, 2016.

Holanda, J. S.; Amorim, J. R. A.; Ferreira Neto, M.; Holanda, A. C. Quality of water for irrigation. In: GHEYI, H. R.; DIAS, N. S.; LACERDA, C. F (ed). Salinity management in agriculture: Basic and applied studies. Fortaleza, INCTA Sal, p. 530, 2016.

Lima, G. S. de; Nobre, R. G.; Gheyi, H. R.; Soares, L. A. A. dos; Silva, A. O. da. Growth and production components of papaya under salt stress and nitrogen fertilization. Engenharia Agricola, v. 34, p.854-866, 2014.

Lima, L. A.; Oliveira, F. A. de; Alves. R. E. C.; Linhares, P. S. F.; Medeiros, A. M. A. de.; Bezerra, F. M. S. Eggplant tolerance to irrigation water salinity. Revista Agroambiente, v.9, p.27-34, 2015.

Kilic, S.; Kahraman, A. The mitigation effects of exogenous hydrogen peroxide when alleviating seed germination and seedling growth inhibition on salinity-induced stress in barley. Polish Journal of Environmental Studies, v. 25, 2016.

Mantovani, A. A method to improve leaf succulence quantification. Brazilian Archives of Biology and Technology, v.42, p.9-14, 1999.

Martinez, J.P.; Lutts, S.; Schanck, A.; Bajji, M.; Kinet, J. M. Is osmotic adjustment required for water stress resistance in the Mediterranean shrub *Atriplex halimus* L. Journal of Plant Physiology, v.161, p.1041-1051, 2004.

Mendonça, V.; Ramos, J. D.; Pio, R.; Gontijo, T. C. A.; Tosta, M. S. Overcoming dormancy and sowing depth of soursop seeds. Revista Caatinga, v. 20, p. 73-78, 2007.

Medeiros, J. F. de. Irrigation water quality and salinity evolution in properties assisted by GAT in the states of RN, PB and CE. (Master's dissertation). Federal University of Paraiba, Campina Grande. 2003, 173p.

Miller, I. M.; Jénsen, P. E.; Hansson, A. Oxidative modifications to cellular components in plants. Annual Review of Plant Biology, v.58, p.459-481, 2007.

Nguyen, G. N.; Hailstones, D. L.; Wilkes, M.; Sutton, B. G. Drought-induced oxidative conditions in rice anthers leading to a programmed cell death and pollen abortion. Journal of Agronomy and Crop Science, v.195, p.157- 164, 2009.

Nobre, R. G.; Lima, G. S. de; Gheyi, H. R., Soares, L. A. A. dos; Silva, A. O. da. Growth, consumption and water use efficiency of papaya under salt and nitrogen stress. Revista Caatinga, v.27, p.148-158, 2014.

Novais, R. F.; Neves J. C. L.; Barros N. F. Controlled environment trial. In: OLIVEIRA A. J. (ed) Métodos de pesquisa em fertilidade do solo. Brasilia: Embrapa-SEA. p. 189-253. 1991.

Oliveira Junior, L D de. Pre-germination treatment of forest seeds with hydrogen peroxide (Master's thesis). Federal University of Lavras, Lavras. 2017, 173p..

Oliveira, M. G. Effect of leaf pretreatment with H2O2 on the proteome and antioxidant enzymes in string bean plants subjected to salt stress (Doctoral thesis.). Fortaleza: UFC, 2016. 128 p.

Petrov, V. D.; Breusegem, F. V. Hydrogen peroxide: A central hub for information flow in plant cells. AoB Plants. v. 2012, p.1-13, 2012

Richards, L. A. Diagnosis and improvement of saline and alkali soils, Washington: U.S, Department of Agriculture, 1954. 160p.

Rutschow, H. L.; Baskin, T. I.; Kramer, E. M. Regulation of solute flux through plasmodesmata in the root meristem. Plant Physiology, v.155, p.1817-1826, 2011.

Sà, F. V. S. da; Brito, M. E. B.; Pereira, I. B.; Neto, P. A.; Andrade Silva, L. de; Costa, F. B. da. Salt balance and initial growth of pine seedlings (*Annona squamosa* L.) under substrates irrigated with saline water. Irriga, v. 20, p. 544, 2015.

Silva, E. N. da.; Silveira, J. A. G.; Rodrigues, C. R. F.; Lima, C. S. de.; Viégas, R. A. Contribution of organic and inorganic solutes in the osmotic adjustment of jatropha submitted to salinity. Pesquisa Agropecuària Brasileira, v.44, p.437-445, 2009.

Silva, E. M. da; Lacerda, F. H. D.; Medeiros, S. A. de; Souza, L. P. de; Pereira, F. H. F. Application methods of different concentrations of H_2O_2 in maize under saline stress. Green Journal of Agroecology and Sustainable Development, v.11, p. 01-07, 2016.

Tanou, G.; Job, C.; Rajjou, L.; Arc, E.; Belghazi, M.; Diamantidis, G.; Uchida, A.; Jagendorf, A. T.; Hibino, T.; Takabe, T. Effects of hydrogen peroxide and nitric oxide on both salt and heat stress tolerance in rice. Plant Science, v. 163, p. 515523, 2002.

Uchida, Akio et al. Effects of hydrogen peroxide and nitric oxide on both salt and heat stress tolerance in rice. Plant Science, v. 163, p. 515-523, 2002.

Wahid, A.; Perveen, M.; Gelani, S.; Basra, S. M. Pretreatment of seed with H2O2 improves salt tolerance of wheat seedlings by alleviation of oxidative damage and expression of stress proteins. Journal of Plant Physiology, v.164, p.283-294, 2007.

CHAPTER III

GAS EXCHANGE AND PHOTOSYNTHETIC PIGMENTS OF SOURSOP UNDER SALT STRESS AND EXOGENOUS H_2O_2 APPLICATION

ABSTRACT: The aim of this study was to evaluate the gas exchange and photosynthetic pigments of Morada Nova soursop seedlings irrigated with saline water and subjected to exogenous application of hydrogen peroxide via seed soaking and foliar spraying. The study was conducted in plastic bags under greenhouse conditions at the Center for Technology and Natural Resources of the Federal University of Campina Grande, PB, using Eutrophic Regolithic Neosol with a sandy loam texture. The treatments were distributed in a randomized block design, in a 5 x 5 factorial arrangement, with five levels of electrical conductivity of the irrigation water - ECa (0.7; 1.4; 2.1; 2.8 and 3.5 dS m^{-1}) and five concentrations of hydrogen peroxide - H_2O_2 (0, 25, 50, 75 and 100 µM), with four replications and three plants per plot. As salt stress increased, there was a decrease in internal CO_2 concentration, instantaneous carboxylation efficiency and water use efficiency, with instantaneous carboxylation efficiency being the most sensitive variable. Hydrogen peroxide at concentrations of 25 and 50 µM attenuated the deleterious effects of water salinity on stomatal conductance, CO_2 assimilation rate and chlorophyll a content, with the 25 µM concentration being the most efficient.

Keywords: *Annona muricata* L., saline water, physiology

1. INTRODUCTION

The soursop is a fruit tree that has stood out for its potential for commercialization in the domestic market with relevant economic importance and prospects for export, with the Northeast region being the largest producer (Braga Sobrinho, 2010; Cavalcante et al., 2017). The consumption of graviola has increased, whether fresh or processed, due to its nutritional importance and the ways in which it can be used in human nutrition, as well as the medicinal properties of its leaves, fruit, seeds and roots (Freitas et al., 2013).

In the semi-arid region of northeastern Brazil, the low levels of rainfall distributed irregularly throughout the year are limiting factors for agricultural production, and the practice of irrigation is the only way to guarantee safe cultivation (Lacerda et al., 2016). However, the water coming from the springs in this region is often saline, which can cause morphological, structural and metabolic changes in plants (Lima et al., 2016). It should also be noted that the effect of water salinity on crops varies between species (Ayres & Westcot, 1999; Brito et al., 2014).

In view of this, research has been carried out using saline water for cultivation in the northeast region, for example, in pine (Sà et al., 2015), citrus (Barbosa et al., 2017) and guava (Sena et al., 2017). It is therefore extremely important to develop research aimed at studying other fruit trees such as soursop, since there are few studies involving the use of saline water in the cultivation of this fruit tree.

In this context, alternatives have been sought to mitigate the effects of salt stress on crops, including the exogenous application of hydrogen peroxide (H O_{22}), in the form of sprays and/or in the pre-treatment of seeds at low concentrations, which has shown promise in acclimatizing crops to salt stress (Gondim et al., 2011). H2O2 is a reactive oxygen species (ROS) capable of oxidizing membrane lipids, denaturing proteins and reacting with DNA, causing mutations (Scandalios, 2002). In addition, it functions as a signaling molecule in plants under biotic and abiotic stress, being involved in various processes such as root gravitropism, tolerance to oxygen deficiency, cell wall strengthening, senescence, photosynthesis, stomatal opening and cell cycle control (Gechev et al., 2006; Petrov & Breusegem, 2012).

The aim of this study was to evaluate the gas exchange and photosynthetic pigments of Morada Nova soursop seedlings irrigated with saline water and exogenous application of hydrogen peroxide.

2. MATERIAL AND METHODS

The work was carried out from May to October 2017, in plastic bags measuring 2 dm3, under greenhouse conditions, belonging to the Center for Technology and Natural Resources of the Federal University of Campina Grande (CTRN/UFCG), located in the municipality of Campina Grande, PB, with geographic coordinates of 07° 15' 18" S latitude, 35° 52' 28" W longitude and an average altitude of 550 m.

The treatments resulted from the combination of five levels of electrical conductivity of the irrigation water - ECa (0.7; 1.4; 2.1; 2.8 and 3.5 dS m^{-1}) and five concentrations of hydrogen peroxide - H O$_{22}$ (0, 25, 50, 75 and 100 µM), distributed in a randomized block design, in a 5 x 5 factorial arrangement, with four replications and three plants per plot, making a total of three hundred plants.

The levels of electrical conductivity of the irrigation water (1.4; 2.1; 2.8 and 3.5 dS m^{-1}) were prepared by dissolving the salts NaCl, CaCl2.2H2O and MgCl2.6H2O, in the equivalent ratio of 7:2:1, respectively, in local supply water (ECa = 1.10 dS m^{-1}). This ratio is commonly found in water sources used for irrigation in small properties in the Northeast (Medeiros et al., 2003), based on the relationship between ECa and salt concentration (10*mmolc L^{-1} = ECa dS m^{-1}) taken from Richards (1954). The level of 0.7 dS m^{-1} was obtained by diluting the local water supply with rainwater (ECa = 0.02 dS m)$.^{-1}$

The plastic bags were filled with 2.6 kg of a substrate made up of soil (84%) + sand (15%) + humus (1%). The soil used in the experiment was Neossolo Regolitico Eutròfico with a sandy loam texture, collected at a depth of 0-20 cm from the rural area of the municipality of Lagoa Seca, PB. The soil was duly pulverized and sieved and its physical, hydraulic and chemical characteristics (Table 1) were determined according to the methodology proposed by Donagema et al. (2011).

Table 3. Chemical and physico-hydric attributes of the soil used in the experiment, before the treatments were applied.

Chemical characteristics									
pH (H2O) (1:2, 5)	M.O. %	P (mg kg)$^{-1}$	K$^+$	In$^+$	Ca^{2+}	Mg^{2+}	Al^{3+} + H$^+$ ......(cmolc kg)$^{-1}$	PST (%)	ECes (dS m)$^{-1}$
5,90	1,36	6,80	2,22	1,60	26,00	36,60	19,30	1,87	1,0

Characteristic			physical-hydraulic conditions					
Gr. fraction	annulometric$_1$:a (dag kg)$^.$ Texture	Humidity	e(kPa)	AD	Porosity		DA	DP

			class 1	33,42	1519.5	total	
Sand	Silt	Clay	....	dag kg^{-1}	....	%	(kg dm)$^{-3}$

O.M. - Organic matter:Walkley-Black wet digestion; Ca^2 + and Mg^2 + extracted with KCl 1 M pH 7.0; Na+ and K^+ extracted using NH4OAc 1 MpH 7.0; Al^3 + and H+ extracted with calcium acetate 1 M pH 7.0; PST - Percentage of exchangeable sodium; ECes - Electrical conductivity of the saturation extract; FA - Sandy loam; AD - Available water; DA - Apparent density; DP - Particle density.

The seeds used in the experiment were obtained from fruit harvested from a commercial orchard located in the municipality of Macaparana, PE. The seeds were extracted manually, then air-dried and the process of breaking dormancy was carried out by cutting them distal to the embryo, according to the methodology proposed by Mendonça et al. (2007).

Before sowing, the seeds were pre-treated with hydrogen peroxide, where they were soaked according to the pre-established concentrations for a period of 24 hours. The seeds were then sown by placing three soursop seeds cv. Morada Nova three centimeters deep and distributed equidistantly; 20 days after germination, the seeds were thinned to obtain only one plant per bag, leaving the one with the greatest vigor.

Before sowing, the soil moisture content was raised to field capacity using water according to the treatment. After sowing, irrigation was carried out daily by applying a volume of water to each plastic bag in order to maintain soil moisture close to field capacity, with the volume applied determined according to the plants' water needs, estimated by the water balance by subtracting the drained volume from the volume applied in the previous irrigation, plus a leaching fraction of 0.10 to control excessive salt accumulation in the root zone, the leaching fraction applied every 20 days.

Fertilization with nitrogen, potassium and phosphorus was carried out as a top dressing, based on the methodology contained in Novais et al. (1991). 0.58 g of urea, 0.65 g of potassium chloride and 1.56 g of monoammonium phosphate were applied, equivalent to 100, 150 and 300 mg kg^{-1} of the substrate of N, K2O and P2O5, respectively, applied as top dressing in four equal applications via fertigation, at 15-day intervals, with the first application made 15 days after sowing (DAS).In order to make up for micronutrient deficiencies, 2.5 g L^{-1} of ubyfol was applied [(N (15%); P2O5 (15%); K2O (15%); Ca (1%); Mg (1.4%); S (2.7%); Zn (0.5%); B (0.05%); Fe

(0.5%); Mn (0.05%); Cu (0.5%); Mo (0.02%)] were applied foliarly at 60 and 100 DAS.

The foliar application of H_2O_2 was done manually at 5 p.m., in the appropriate concentrations, at 90, 105 and 120 DAS, spraying the abaxial and adaxial sides of the leaves so that the foliage was completely wet, using a spray bottle.

Gas exchange was measured at 120 days after sowing (DAS) through stomatal conductance (mol of H_2O m^{-2} s^{-1}), transpiration (mmol of H_2O m^{-2} s^{-1}), CO_2 assimilation rate (μmol m^{-2} s^{-1}) and internal CO_2 concentration (μmol m $s^{-2 -1}$) (IC) were evaluated on the third leaf, counted from the apex, using the portable photosynthesis measuring equipment "LCPro+" from ADC BioScientific Ltda. These data were used to quantify the instantaneous efficiency of water use (EIUA) (A/E) [(μmol m^{-2} s^{-1}) (mol H_2O m^{-2} s)$^{-1-1}$] and the instantaneous efficiency of carboxylation (EICI) (A/Ci) [(μmol m^{-2} s^{-1}) (μmol mol]$^{-1-1}$ (Konrad et al., 2005; Jaimez et al., 2005).

Quantification of photosynthetic pigment levels (chlorophyll a and b and carotenoids) was carried out using the laboratory method developed by Arnon (1949), where plant extracts were made from samples of discs from the limbus of the third mature leaf from the apex. From these extracts, the concentrations of chlorophyll and carotenoids in the solutions were determined using a spectrophotometer at absorbance wavelengths (ABS) (470, 646, and 663 nm), using the following equations: Chlorophyll a (Cla) = (12.21 x ABS663) - (2.81 x ABS646); Chlorophyll b (Clb) = (20.13 x ABS64) - (5.03 x ABS663) and Carotenoids (Car) = ((1000 x ABS470) - (1.82 x Cla) - (85.02 x Clb))/198. The values obtained for the chlorophyll a, b and carotenoid contents in the leaves were expressed in μm g^{-1} MF (fresh matter).

The data collected was subjected to analysis of variance using the F test at 0.05 and 0.01 probability levels and, when significant, linear and quadratic polynomial regression analysis was carried out using the SISVAR statistical software (Ferreira, 2014).

3. RESULTS AND DISCUSSION

Based on the summary of the F test (Table 2), it can be seen that the salinity levels of the irrigation water significantly affected stomatal conductance (gs), transpiration (E), CO assimilation rate$_2$ (A), internal CO concentration$_2$ (CI), instantaneous carboxylation efficiency (EICI) and instantaneous water use efficiency (EIUA). There was a significant effect of hydrogen peroxide concentrations and the interaction between the factors (salinity levels - NS x hydrogen peroxide - H_2O_2) for gs, E and A.

Table 2. Summary of the F test for stomatal conductance *(gs)*, transpiration (E), CO_2 assimilation rate (A), internal CO_2 concentration (IC), instantaneous carboxylation efficiency (ICIE) and instantaneous water use efficiency (WUE) of Morada Nova soursop plants irrigated with saline water and subjected to exogenous application of hydrogen peroxide at 120 days after sowing.

Source of variation	F-test					
	gs	E	A	CI	EICI[1]	EIUA
Salt Levels (NS)	**	**	**	*	**	**
Linear regression	**	**	**	**	**	**
Quadratic regression	ns	ns	ns	ns	ns	ns
Hydrogen Peroxide (H_2O_2)	**	**	**	ns	ns	ns
Linear regression	**	*	*	ns	ns	ns
Quadratic regression	ns	**	**	ns	ns	ns
Interaction (NS x H_2O_2)	**	**	**	ns	ns	ns
Blocks	ns	ns	ns	ns	ns	ns
CV (%)	19,99	14,66	17,88	18,94	25,30	23,45

[ns], **, * respectively not significant, significant at $p < 0.01$ and $p < 0.05$

The increase in the electrical conductivity of the irrigation water negatively affected the stomatal conductance of the Morada Nova soursop plants in the control treatment, i.e. those that did not receive exogenous application of H_2O_2; according to the regression equation (Figure 1A), a linear effect was observed, with a decrease of around 13.64% per unit increase in ECa. When comparing in relative terms the results obtained from the plants subjected to the highest level of salinity (3.5 dS m-1) in relation to the lowest level (0.7 dS m-1), there was a 42.23% decrease in stomatal conductance. However, it can be seen that the concentrations of 25 and 50 μM of hydrogen peroxide were able to attenuate the harmful effects of water salinity from the saline level of 1.4 dS m^{-1} , with the concentration of 25 μMa being more efficient. Concentrations of 75 and 100 μM intensified the negative effects of salinity.

The increase in stomatal conductance of Morada Nova soursop plants subjected to a hydrogen peroxide concentration of 25 µM indicates a recovery in stomatal movement, possibly signaled by exogenous H O_{22} . Pre-exposure of plants to moderate stresses or to signaling metabolites, such as $H2O2$, can result in metabolic signaling in the cell (increased antioxidative metabolites and/or enzymes) and, therefore, better physiological performance when the plant is exposed to more severe stress conditions (Veal et al., 2007; Forman et al., 2010).

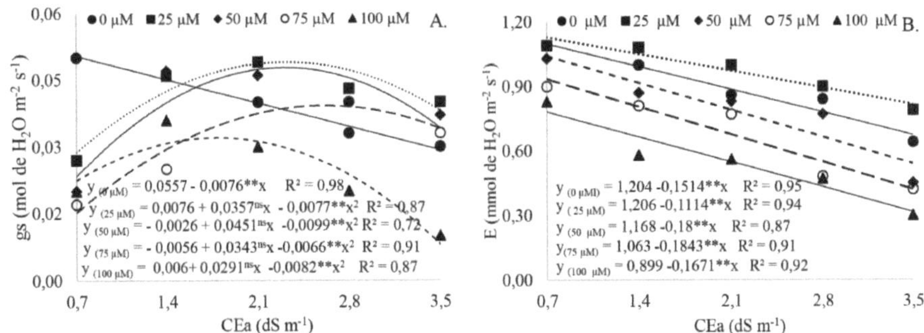

Figura 1. Stomatal conductance - gs (A) and transpiration - E (B) of soursop plants cv. Morada Nova, as a function of the interaction between the electrical conductivity of the irrigation water - ECa and the concentrations of hydrogen peroxide

When studying the interaction between the salinity of the irrigation water and the concentrations of hydrogen peroxide on the transpiration of soursop cv. According to the regression equation (Figure 1B), the concentration of 25 µM of H2O2 promoted greater transpiration when compared to the values for plants in the control treatment (0 µM); in relation to the other concentrations of H2O2, it can be seen that they intensified the deleterious effects of water salinity. The decreases per unit increase in ECa were 12.57, 9.23, 15.41, 17.33 and 18.58% respectively for the concentrations of 0.25, 50, 75 and 100 µM.

The rate of CO2 assimilation was reduced linearly as the electrical conductivity of the irrigation water increased. According to the regression equation (Figure 2A) in the control treatment plants (0 µM), there was a linear decrease of 21.79% per unit increase in ECa, i.e. a 72.01% reduction in the A of plants irrigated with the highest salinity water (3.5 dS m^{-1}) in relation to the lowest saline level (0.7 dS m^{-1}). However, there was a 138% increase in the rate of CO_2 assimilation in the Morada Nova grapevine plants subjected to an H_2O_2 concentration of 25 µM and irrigated with water of 3.5 dS m^{-1} when compared to the control treatment (0.7 dS m^{-1}). As observed in the gs and E variables (Figures 1A and B), the application of different concentrations of hydrogen peroxide did not attenuate the deleterious effect of salinity on the CO_2 assimilation rate of the Morada Nova grapevine, the opposite occurring; thus, it can be inferred that the excess of reactive oxygen species has a toxic effect, caused above all by oxidative stress.

Salt stress reduced stomatal conductance, transpiration and the rate of CO_2 assimilation in Morada Nova soursop plants in the control treatment (0.7 dS m^{-1}). Stomatal closure in plants has the effect of restricting the entry of CO_2 into the cells of the leaf mesophyll, a fact that can increase susceptibility to photochemical damage, since lowering the rate of CO assimilation$_2$ causes excessive light energy in photosystem II (Munns & Tester, 2008; Silva et al., 2010).

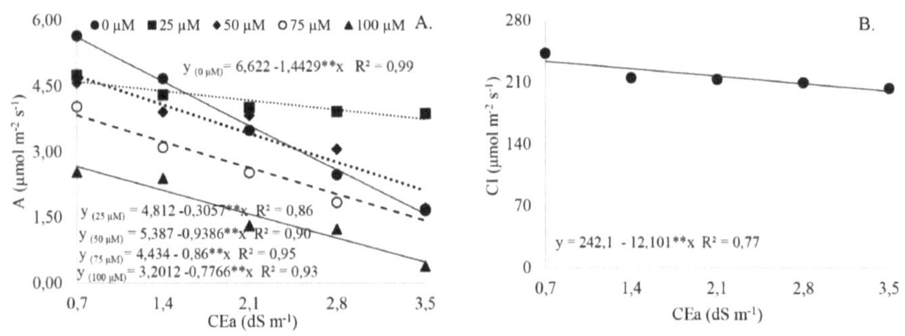

Figura 2. The CO assimilation rate$_2$ - A (A) of the soursop tree as a function of the interaction between the electrical conductivity of the irrigation water and hydrogen peroxide concentrations and the internal CO_2 concentration - CI (B) as a function of water salinity

The beneficial effect of hydrogen peroxide at low concentrations may be associated with its role as a signaling molecule, regulating various pathways, including responses to salt stress (Baxter et al., 2014). Therefore, H2O2 is related to the regulation of various mechanisms under abiotic and biotic stress conditions (Orozco Càrdenas et al., 2001; Malolepsza & Rôzÿalska, 2005).

The internal CO_2 concentration (IC) of the soursop plants cv. According to the regression equation (Figure 2B), at 1.4 dS m^{-1} there was a reduction of 3.62% compared to the control treatment (0.7 dS m^{-1}); for the ECa values of 2.1, 2.8 and 3.5 dS m^{-1} the reductions were 7.25, 10.87 and 14.5%, respectively. The reductions in internal CO_2 concentration with increasing salt levels recorded in the Morada Nova grapevine is a common response of plants to salt stress, which is probably due to the lower diffusion of $CO_{2\ in}$ the substomatal chamber as a result of the stomata closing (Silva et al., 2011; Oliveira et al., 2017).

With regard to instantaneous carboxylation efficiency (ICEE), there was a negative effect of irrigation water salinity and, according to the regression equation (Figure 3A), there was a reduction in ICEE of 18,49% per unit increase in the electrical conductivity of the water, i.e. a 59.47% reduction in the EICI in plants irrigated with the highest salinity water (3.5 dS m^{-1}) compared to the lowest salinity level (0.7 dS m^{-1}). According to Taiz & Zeiger (2017), as the stress becomes severe, the dehydration of the mesophyll cells inhibits photosynthesis, thus damaging the metabolism and, consequently, compromising the efficiency of carboxylation. According to Lacher (2006), this reduction in EICI is related to metabolic restrictions in the Calvin cycle, where the carbon received is not used in the carboxylation stage of the mesophyll cells.

The instantaneous efficiency of water use (EIUA) was also negatively affected by the salinity of the irrigation water and the regression equation (Figure 3B) shows that the plants irrigated with water with an ECa of 0.7 dS m^{-1} (control treatment) had the highest EIUA [4,48 µmol m^{-2} s^{-1} (mol $_{H2O}$ m^{-2} s)$^{-1-1}$] and the lowest [3.37 µmol m^{-2} s^{-1} (mol $_{H2O}$ m s)$^{-2-1-1}$] with the plants grown with water with the highest saline level (3.5 dS $_{m-1}$), i.e., a reduction of 24.74% [1.11 µmol m^{-2} s^{-1} (mol $_{H2O}$ m^{-2} s)$^{-1-1}$] between the highest (3.5 dS m^{-1}) and lowest (0.7 dS $_{m-1}$) salinity levels of the irrigation water. It can be inferred that increasing the salinity of irrigation water directly affects the USE of Morada Nova soursop plants. Thus, the reduction in USE observed in this research may be related to the accumulation of salts in the soil throughout the crop cycle, a situation which contributed to a reduction in the soil's osmotic potential and, consequently, made it more difficult for the plants to absorb water (Nobre et al., 2014).

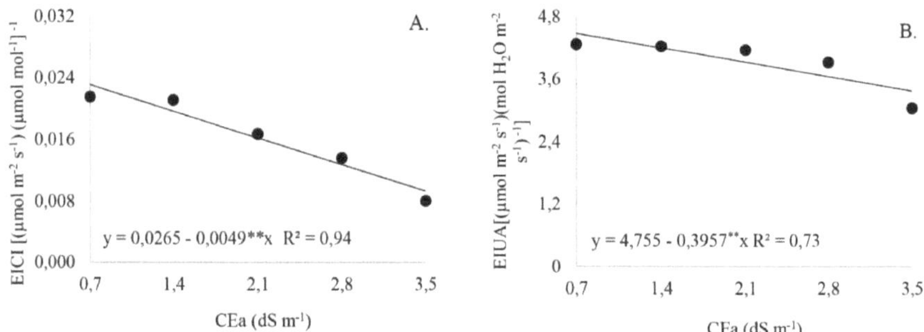

Figura 3. Instantaneous carboxylation efficiency - EICI (A) and instantaneous water use efficiency - EUIA (B) of soursop as a function of irrigation water salinity The F test (Table 3) showed a significant effect ($p < 0.01$) of irrigation water salinity levels on chlorophyll a (Cla) and chlorophyll b (Clb). The concentrations of hydrogen peroxide and the interaction between the factors (NS x H O_{22}) had a significant influence ($p < 0.01$) on all the variables analyzed.

Table 3. Summary of the F test for chlorophyll a (Cla), chlorophyll b (Clb) and carotenoids (Car) in Morada Nova soursop plants irrigated with saline water and subjected to exogenous hydrogen peroxide at 140 days after sowing.

Variation source	F-test		
	Cla	Clb	Car
Salt Levels (NS)	**	**	ns
Linear regression	**	**	ns
Quadratic regression	ns	ns	ns
Hydrogen peroxide (H2O2)	**	**	**
Linear regression	ns	ns	*
Quadratic regression	**	**	ns
Interaction (NS x H2O2)	**	**	**
Blocks	ns	ns	ns
CV (%)	8,11	15,69	14,87

ns, **, * respectively not significant, significant at p < 0.01 and p < 0.05

The regression equation (Figure 4A) shows that the plants in the control treatment (0 µM) had a reduction in Cla content of 15.2% per unit increase in the electrical conductivity of the irrigation water. On the other hand, Morada Nova soursop plants treated with hydrogen peroxide at concentrations of 25 and 50 µM had the deleterious effects of salinity on chlorophyll a attenuated. The graviola plants cv. Morada Nova irrigated with the highest saline level (3.5 dS m^{-1}) obtained the highest averages for Cla when subjected to 50 µM of H O$_{22}$ (4.92 µm g^{-1} MF), representing an increase of 47.15% compared to the control treatment (0 µM). However, the concentrations of 75 and 100 µM did not mitigate the negative effects of irrigation water salinity on the Cla content of Morada Nova graviola.

The chlorophyll b (Clb) of Morada Nova soursop plants was significantly influenced by the interaction between salt level - NS x H2O2. Through the regression equations (Figure 4B), it can be seen that the treatment with the H2O2 concentration of 50 µM promoted the highest averages for the chlorophyll b content corresponding to 1.59; 1.78 and 1.66 µm g^{-1} MF, when irrigated, respectively, with the ECa levels of 2.1; 2.8 and 3.5 dS m^{-1}. It can also be seen (Figure 4B) that the plants in the control treatment (0 µM) had a 13.87% reduction in chlorophyll b content per unit increase in ECa. However, the Morada Nova soursop plants subjected to a concentration of 50 µM and exposed to salt stress at the highest level of ECa (3.5 dS m^{-1}), obtained an increase of 0.875 µm g^{-1} MF in the chlorophyll b content compared to those subjected to the control treatment. The decrease in chlorophyll b content observed in the plants of the control treatment (0 µM) can be attributed to an increase in the activity of the enzyme chlorophyllase, which degrades chlorophyll, given that salt stress induces the degradation of β-carotene and a reduction in the formation of zeaxanthin, producing a decrease in the content of carotenoids, pigments apparently involved in protection against photoinhibition (FREIRE et al., 2013). Godim (2012), evaluating leaf pre-treatment with H2O2 as a strategy to minimize the deleterious effects of salinity on maize plants, observed that the highest chlorophyll contents were obtained in plants pre-treated with H2O2 at a concentration of 10 mM even when subjected to saline stress (80 mM NaCl).

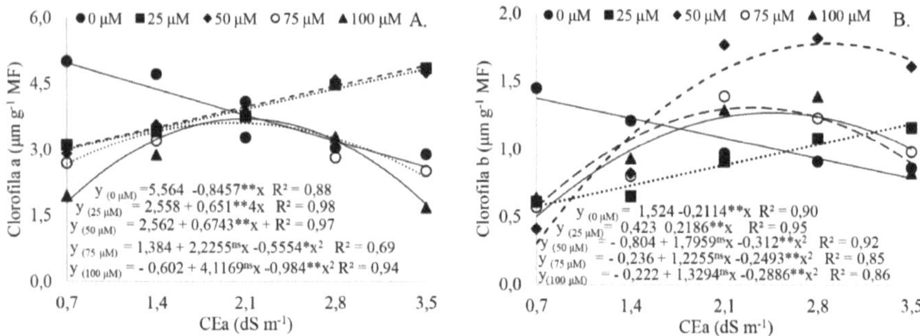

Figura 4. Chlorophyll a (A) and Chlorophyll b (B) of soursop cv. Morada Nova as a function of the interaction between electrical conductivity of irrigation water and concentrations of hydrogen peroxide

As far as carotenoid content is concerned, according to the regression equation (Figure 5), there was a linear and decreasing effect on plants in the control treatment (0 μM), with a decrease in carotenoid content of 14. 55% per unit increase in ECa (0.844 μm g MF),55% per unit increase in ECa, i.e. a reduction of 45.37% (0.844 μm g⁻¹ MF) in the carotenoid content of plants irrigated with water of 3.5 dS m⁻¹ compared to the lowest level of 0.7 dS m⁻¹ . However, it was noticed that the plants submitted to treatment with hydrogen peroxide had an increase in carotenoid content even with the imposition of salt stress, especially at a concentration of 25 μM, in which the regression equation (Figure 5) shows an increase of 45,46% (0.847 μm g⁻¹ MF) compared to the control treatment when the plants were irrigated with the water with the highest salt level (3.5 dS m⁻¹), indicating that hydrogen peroxide was effective in acclimatizing soursop plants cv. Morada Nova to salt stress. Carotenoids are pigments that can exert a photoprotective action on the photochemical apparatus, and the increase in carotenoids is possibly a defense mechanism, preventing photo-oxidative damage to chlorophyll molecules (Kerbauy, 2004; Raven et al., 2007).

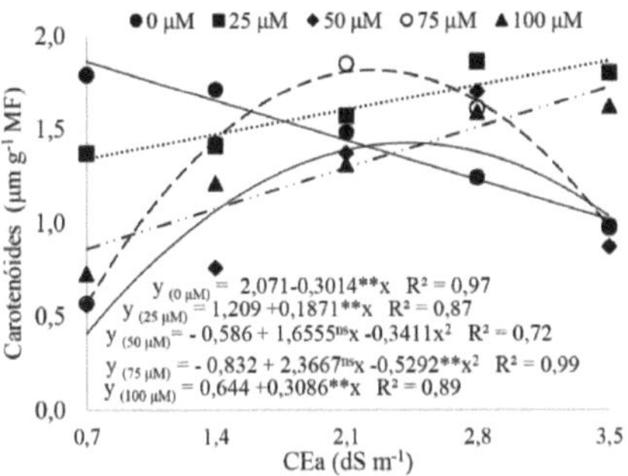

Figura 5. Carotenoids of soursop cv. Morada Nova, as a function of the interaction between electrical conductivity of irrigation water - ECa and hydrogen peroxide concentrations

The reduction in carotenoid content in the control treatment may possibly have occurred as a result of the degradation or inhibition of carotenoid synthesis, mainly due to photo-oxidation, causing damage to photosynthetic membranes, as well as affecting other cellular processes such as cell division and expansion (Silva et al., 2014).

4. CONCLUSIONS

As salt stress increases, there is a decrease in the photosynthetic parameters of Morada Nova soursop plants.

Exogenous application of hydrogen peroxide at concentrations of 25 and 50 μM attenuated the deleterious effects of salt stress on stomatal conductance, CO_2 assimilation rate and chlorophyll a content.

Hydrogen peroxide at concentrations of 75 and 100 μM in interaction with irrigation water salinity promotes a decrease in transpiration, CO_2 assimilation rate and chlorophyll a content in Morada Nova soursop.

5. BIBLIOGRAPHICAL REFERENCES

Ayers, R. S.; Westcot, D. W. 1999. Water quality in agriculture. Campina Grande: UFPB. 184p. FAO Irrigation and Drainage Studies, 29.

Arnon, D. I. Copper enzymes in isolated chloroplasts: polyphenoloxidases in *Beta vulgaris*. Plant Physiology, v.24, p.1-15, 1949.

Barbosa, R. C. A.; Brito, M. E. B.; Sà V. S, F.; Soares Filho, W. S.; Fernandes, P. D.; SILVA, L. A. Gas exchange of citrus rootstocks in response to intensity and duration of saline stress. Semina: Agricultural Sciences, v. *38*, 2017.

Baxter, A.; Mittler, R.; Suzuki, N. EROS as key players in plant stress signaling. Journal of Experimental Botany, v. 65, p. 1229-1240, 2014.

Braga Sobrinho, R. Potential exploitation of anonaceae in the Northeast of Brazil. In: SEMANA DA FLORICULTURA E AGROINDÛSTRIA, 17, 2010, Fortaleza. Proceedings... Fortaleza: Embrapa Agroindustria Tropical, 2010.

Brito, M. E. B.; Fernandes, P.D.; Gheyi, H.R.; Melo, A.S.; Soares Filho, W.S.; Santos, R.T. Sensitivity to salinity of trifoliate hybrids and other citrus rootstocks. Revista Caatinga, v. 27, p. 17-27, 2014.

Cavalcante, L. F.; Da Rocha, L. F.; Silva, R. A. R.; Souto, A. G. L.; Nunes, J. C.; Cavalcante, í. H. L. Production and quality of graviola under irrigation and soil cover with sisal residue. Magistra, v. *28*, p.91-101, 2017.

Donagema, G. K.; Campos, D. V. B. de; Calderano, S. B.; Teixeira, W. G.; Viana, J. H. M. (Org.). Manual of soil analysis methods. 2.ed. Rio de Janeiro: Embrapa Solos, 2011.

Ferreira, D. F. Sisvar: A guide for its bootstrap procedures in multiple comparisons. Ciência e Agrotecnologia, v. 38, p. 109-112, 2014.

Forman, H. J.; Maiorino, M.; Ursini, F. Signaling functions of reactive oxygen species. Biochemistry, v.49, p.835-842, 2010

Freire, J. L. O.; Cavalcante, L. F.; Nascimento, R.; Rebequi, A. M. Chlorophyll content and leaf mineral composition of passion fruit irrigated with saline water and biofertilizer. Revista de Ciências Agrârias, v.36, p.57-70, 2013

Freitas, A. L. G. E.; Vilasboas, F. S.; Pires, M. M.; Sâo José, A. R. Characterization of Graviola (Annona muricata L.) Production and Market in the State of Bahia. Informaçôes Econômicas, v. 43, p.23-34, 2013.

Gechev, T. S.; Van Breusegem, F.; Stone, J. M.; Denev, I.; Laloi, C. Reactive oxygen species as signals that modulate plant stress responses and programmed cell death Bioessays, v.28, p.1091-1101, 2006.

Gondim, F. A.; Gomes Filho, E.; Marques, E. C.; Prisco, J. T. Effects of H2O2 on growth and solute accumulation in maize plants under salt stress. Revista Ciência Agronômica, v. 42, p. 373-38, 2011.

Gondim, F A. Foliar pretreatment with H2O2 as a strategy to minimize the deleterious effects of salinity on corn plants. Federal University of Ceará, Fortaleza. 2012, 147p. PhD Thesis

Jaimez, R. E.; Rada, F.; Garcia-Nûnez, C.; Azócar, A. Seasonal variations in leaf gas exchange of platain cv. Hartón (Musa AAB) under different soil water conditions in a humid tropical region. Scientia Horticulturae, v.104, p.79-89, 2005.

Kerbauy, G.B. Fisiologia vegetal. Rio de Janeiro: Guanabara Koogan, 2004. 452p.

Konrad, M. L. F.; Silva, J. A. B.; Furlani, P. R.; Machado, E. C. Gas exchange and chlorophyll fluorescence in six coffee cultivars under aluminium stress. Bragantia, v.64, p.339-347, 2005.

Lacerda, C. F.; Costa, R. N. T.; Bezerra, M. A.; Neves A. L. R.; Sousa, G. G.; Gheyi, H. R. Management strategies for the use of saline water in agriculture. In: GHEYI, H. R.; DIAS, N. S.; (ed). Salinity management in agriculture: Basic and applied studies. Fortaleza, INCTA Sal, p. 530, 2016.

Larcher, W. Plant ecophysiology. Sao Carlos: Rima Artes e Textos, 2006. 550p.

Lima, G. S. de; Santos, J. B. dos; Soares, L. A. A. dos; GheyI, H. R.; Nobre, R. G.; PEREIRA, R. F. Irrigation with saline water and foliar proline application in 'All Big' pepper cultivation. Comunicata Scientiae,v.7, p.513, 2016.

Malolepsza, U.; Rozalska, S. Nitric oxide and hydrogen peroxide in tomato resistance: Nitric oxide modulates hydrogen peroxide level in o-hydroxyethylorutin-induced resistance to Botrytis cinerea in tomato. Plant Physiology and Biochemistry, France, v. 43, p.623-635, 2005.

Mendonça, V.; Ramos, J. D.; Pio, R.; Gontijo, T. C. A.; Tosta, M. S. Overcoming dormancy and sowing depth of soursop seeds. Revista Caatinga, v. 20, p. 73-78, 2007.

Medeiros, J. F. de. Irrigation water quality and salinity evolution in properties assisted by GAT in the states of RN, PB and CE. Federal University of Paraiba, Campina Grande. 2003, 173p. Master's dissertation

Munns, R.; Tester, M. Mechanisms of salinity tolerance. Annual Review of Plant Biology, v.59, p.651-681, 2008.

Nobre, R. G.; Lima, G. S de; Gheyi, H. R.; Soares, L. A. A.; Silva, A.O. Growth, consumption and efficiency of water use by papaya under saline and nitrogen stress. Revista Caatinga, v.27, p.148-158, 2014.

Novais, R. F.; Neves J. C. L.; Barros N. F. Controlled environment trial. In: OLIVEIRA A. J. (ed) Métodos de pesquisa em fertilidade do solo. Brasilia: Embrapa-SEA. p. 189-253. 1991.

Oliveira, W. J. D.; Souza, E. R. D.; Cunha, J. C.; Silva, Ê. F. D. F.; Veloso, V. D. L. Leaf gas exchange in cowpea and CO2 efflux in soil irrigated with saline water. Revista Brasileira de Engenharia Agricola e Ambiental, v.21, p. 32-37, 2017.

Orozco-Cârdenas, M. L.; Narvâez-Vâsquez, J.; Ryan, C. A. Hydrogen peroxide acts as a second messenger for the induction of defense genes in tomato plants in response to wounding, system in, and methyl jasmonate. The Plant Cell, v.13, p.179-191, 2001.

Petrov, V. D.; Breusegem, F. V. Hydrogen peroxide: A central hub for information flow in plant cells. AoB Plants. v. 2012, p.1-13, 2012

Raven, P.H.; Evert, R.F.; Eichhorn, S.E. Plant biology. 7.ed. Rio de Janeiro: Guanabara Koogan, 2007. 728p.

Richards, L. A. Diagnosis and improvement of saline and alkali soils, Washington: U.S, Department of Agriculture, 1954. 160p.

Sà, F. V. S. da; Brito, M. E. B.; Pereira, I. B.; Neto, P. A.; Silva, L. de A.; Costa, F. B. da. Salt balance and initial growth of pine seedlings (*Annona squamosa* L.) under substrates irrigated with saline water. Irriga, v.20, p. 544, 2015.

Sena, G. S. A.; Nobre, R. G., Souza, L. de P.; Barbosa, J. L.; De Souza, C. M. A.; Elias, J. J. Formation of guava rootstock submitted to different water salinities and nitrogen fertilization. Brazilian Journal of Irrigated Agriculture, v.11, p.1578, 2017.

Scandalios, J. G. The rise of ROS. Biochemical Science, v.27, p.483-486, 2002.

Silva, E. D.; Ribeiro, R. V.; Ferreira-Silva, S. L.; Viégas, R. A.; Silveira, J. A. G. Comparative effects of salinity and water stress on photosynthesis, water relations and growth of Jatropha curcas plants. Journal of Arid Environments, v.74, p.11301137, 2010.

Silva, E. N. da; Ribeiro, R.V.; Ferreira-Silva, S.L.; Viégas, R.A.; Silveira, J.A.G. Salt stress induced damages on the photosynthesis of physic nut young plants. Scientia Agricola, v.68, p.62-68, 2011.

Silva, M. A.; Santos, C. M.; Vitorino, H. S.; Rhein, A. F. L.; Photosynthetic pigments and Spad index as descriptors of water deficiency stress intensity in sugarcane. Bioscience Journal, v.30, p.173-181. 2014.

Taiz, L.; Zeiger, E. Plant physiology. 6. ed. Porto Alegre: Artmed, 2017, 888p.

Veal, E. A.; Day, A. M.; Morgan, B. A. Hydrogen peroxide sensing and signaling. Molecular cell, v.26, p.1-14, 2007.

CHAPTER IV

GROWTH AND QUALITY OF SOURSOP SEEDLINGS UNDER SALINE STRESS AND EXOGENOUS APPLICATION OF H2O2

ABSTRACT: The aim of this study was to evaluate the growth and quality of Morada Nova soursop seedlings as a function of irrigation with saline water and exogenous applications of hydrogen peroxide (H_2O_2), under greenhouse conditions at the Center for Technology and Natural Resources of the Federal University of Campina Grande, PB. The experimental design used was randomized blocks, in a 5 x 5 factorial scheme, consisting of a combination of five levels of electrical conductivity of the irrigation water - ECa (0.7; 1.4; 2.1; 2.8 and 3.5 dS m^{-1}) and five concentrations of hydrogen peroxide (0; 25; 50; 75 and 100 µM). The hydrogen peroxide concentrations were applied by soaking the seeds for a period of 24 hours and via foliar application by spraying the leaves thoroughly. Irrigation with water with an electrical conductivity of 0.7 dS m and above^{-1} negatively affected the growth and quality of Morada Nova soursop seedlings, with root dry mass being the most sensitive variable to salt stress. Concentrations of 31 and 100 µM of hydrogen peroxide promoted higher values of relative growth rate in leaf area and leaf and stem dry matter, respectively. The use of saline water with an electrical conductivity of 1.22 dS m^{-1} can be used to irrigate Morada Nova soursop seedlings, as it promotes an acceptable average reduction of up to 10% in growth.

Keywords: *Annona muricata* L., salinity, acclimatization

1. INTRODUCTION

Belonging to the Anonaceae family, the soursop tree (*Annona muricata*, L.) occupies a promising position in Brazilian fruit growing, especially in the Northeast region, where its consumption has increased, whether in natura or industrially processed, due to its nutritional importance and its use in human food, as well as the medicinal properties of its leaves, fruits, seeds and roots (Freitas et al., 2013).

Although the Northeast has favorable soil and climate conditions for the production of soursop, this is not enough to have a great potential for exploiting this crop, since this potential has been limited due to irregular rainfall regimes (Sà et al., 2015), resulting in a water deficit for the plants as the evapotranspiration rate exceeds the precipitation rate for most of the year, causing saline levels in the water sources to rise (Holanda et al., 2016).

As a result, irrigation using water with a high concentration of soluble salts, especially sodium, has become common practice; moreover, the use of this water for a prolonged period of time can have negative effects on the soil and the crops planted in these areas, compromising their growth and development due to toxic, osmotic and nutritional imbalance effects, causing morphological and physiological changes, and consequently compromising production (Neves et al., 2009).

In this way, the formation of soursop seedlings in this region can be optimized with the use of techniques that make it feasible to manage water with excess salts, including the process of acclimatization, which consists of the prior exposure of seeds to a certain type of stress, causing metabolic changes that are responsible for increasing their tolerance to new exposure to stress (Aragâo et al., 2011).

The use of hydrogen peroxide (H_2O_2) exogenously and in low concentrations in plants has emerged as a promising alternative for acclimatizing plants to salt stress, since this molecule generally improves the capacity of the plant's antioxidant system, which, rapidly acts on the reactive oxygen species (ROS) produced by the stress, neutralizing their action or preventing their generation, resulting in a lower concentration of ROS and, consequently, causing less cell damage (iseri et al., 2013).

Pretreatment of seeds with hydrogen peroxide can also act by increasing stomatal conductance, photosynthesis, chlorophyll levels and the protection of chloroplast membranes; as a result, plants in general show greater growth and accumulation of dry mass (Ahmad et al., 2013). However, there is little information on seed pre-treatment and foliar application of H2O2 to soursop and its effects on tolerance to salt stress. In this sense, studies that make it feasible to use hydrogen peroxide to acclimatize soursop plants to salt stress are important for the full development of this crop in the semi-arid region of northeastern Brazil.

The aim of this study was to evaluate the effects of exogenous application of different concentrations of hydrogen peroxide on the growth and quality of Morada Nova soursop seedlings, irrigated with water of different saline levels.

2.   MATERIAL AND METHODS

The experiment was conducted between May and October 2017, in a greenhouse at the Center for Technology and Natural Resources of the Federal University of Campina Grande (CTRN/UFCG), in the municipality of Campina Grande, PB, located at the geographic coordinates 7° 15'18" South latitude and 35° 52' 28" West longitude of the Greenwich Meridian and at an altitude of 550m.

The experimental design used was entirely randomized blocks in a 5 x 5 factorial scheme, with four replications, referring to the levels of salinity of the irrigation water, expressed by the values of electrical conductivity - ECa (0.7; 1.4; 2.1; 2.8 and 3.5 dS m^{-1}), and five concentrations of hydrogen peroxide (0; 25; 50; 75 and 100 μM). The Morada Nova soursop cultivar was used in the experiment because it is more appreciated by producers and makes up the majority of commercial orchards in Brazil, as well as having larger fruit, which can reach a weight of up to 15 kg, and because it produces more than the others (Sacramento et al., 2009).

The seeds used in the experiment were obtained from fruit harvested from a commercial orchard located in the municipality of Macaparana, PE. The seeds were extracted manually and then dried in the shade and air. After drying, the process of breaking dormancy was carried out by cutting off the distal part of the seed, according to the guidelines of Mendonça et al. (2002).

The waters with different saline levels were prepared by adding sodium chloride (NaCl), calcium (CaCl2 .2H2 O) and magnesium (MgCl2 .6H2 O) salts in such a way as to have an equivalent ratio of 7:2:1. This ratio of salts is commonly found in water sources used for irrigation in small properties in the Northeast of Brazil (Medeiros et al... 2003), 2003), with the quantities determined taking into account the relationship between the ECa and the salt concentration ($10*mmolc_{L-1} = ECa - dS\ m\ )^{-1}$

The different concentrations of hydrogen peroxide (H_2O_2), previously established according to the studies, were obtained by diluting $H\ O_{22}$ in deionized water. Before sowing, the seeds were immersed in the hydrogen peroxide solutions following the treatments, for a period of 24 hours in the dark; immediately after this period, sowing took place. Sowing, using 3 seeds for each seedling, was carried out in plastic bags with a capacity of 2 dm^3 of soil, perforated on the sides to allow free drainage. The bags were placed on wooden benches at a height of 0.80 m from the ground and filled with a substrate made up of soil (84%) + sand (15%) + humus (1%) by volume.

The soil used in the experiment was Neossolo Regolitico Eutròfico with a sandy loam texture, collected at a depth of 0-20 cm from the rural area of the municipality of Lagoa Seca, PB. It was duly pulverized and sieved and its physical, hydraulic and chemical characteristics (Table 1) were determined according to the methodology proposed by Donagema et al. (2011).

Table 4. Chemical and physico-hydric attributes of the soil used in the experiment, before the start of the experiment.

Chemical characteristics									
pH (H_2O) (1:2, 5)	M.O. %	P (mg kg)$^{-1}$	K^+	In^+	Ca^{2+}	Mg^{2+}	$Al^{3+} + H^+$	PST (%)	CEes (dS m)$^{-1}$
			...........(cmolckg)$^{-1}$...........						
5,90	1,36	6,80	2,22	1,60	26,00	36,60	19,30	1,87	1,0

Physico-hydric characteristics									
Particle size (dag kg)$^{-1}$			Textural class	Humidity (kPa)		AD	Total porosity %	DA (g cm)$^{-3}$	DP
Sand	Silt	Clay		33,42	1519.5 dag kg^{-1}				
732,9	142,1	125,0	FA	11,98	4,32	7,66	47,74	1,39	2,66

O.M. - Organic matter:Walkley-Black wet digestion; Ca^{2+} and Mg^{2+} extracted with KCl 1 mol L^{-1} pH 7.0; Na^+ and K^+ extracted using NH4OAc 1 mol L^{-1} pH 7.0; Al^{3+} and H^+ extracted with calcium acetate 1 mol L^{-1} pH 7.0; PST- Percentage of exchangeable sodium; ECes - Electrical conductivity of the saturation extract; FA - Sandy loam; AD - Available water; DA- Apparent density; DP- Particle density.

During the research, the soil was kept close to field capacity with daily irrigations, applying the water corresponding to the treatments in each bag, with the volume applied estimated by the water balance: volume of water applied minus the volume of water drained in the previous irrigation, plus a leaching fraction of 0.15, in order to avoid excessive accumulation of salts in the soil.

Nitrogen (N), potassium (K) and phosphate (P) fertilizations were carried out based on Novais et al. (1991). 0.58 g of urea, 0.65 g of potassium chloride and 1.56 g of monoammonium phosphate were applied, equivalent to 100, 150 and 300 mg kg^{-1} of the substrate of N, K_2O and P_2O_5, respectively; applied as top dressing in four equal applications, via fertigation, at 15-day intervals, with the first application made 15 days after sowing (DAS).

The foliar application of H_2O_2 was carried out manually at 5 p.m., in the appropriate concentrations, at 90, 105 and 120 DAS, spraying the abaxial and adaxial sides of the leaves so that the leaves were completely wet, using a spray bottle.

The effects of the treatments were evaluated by determining the relative growth rate in plant height (TCRap), stem diameter (TCRdc) and leaf area (TCRaf), from 45 to 145 DAS, as well as the dry phytomass of leaf (FSF), stem (FSC), root (FSR) and total (FST), and the Dickson quality index (IQD), at 145 DAS.

The relative growth rates (RGR) were estimated from the values of plant height, stem diameter and leaf area and obtained according to the methodology contained in Benincasa (2003) expressed by Equation 1.

$$TCR = \frac{(\ln A_2 - \ln A_1)}{(t_2 - t_1)} \tag{1}$$

In which:

TCR - relative growth rate (mm mm^{-1} day^{-1} ; cm cm^{-1} day)$^{-1}$

A_1 - variable in time t_1 ;

A_2 - variable in time t_2 ; e,

ln - natural logarithm.

To obtain the dry phytomass, the stem of each plant was cut close to the ground and then the different parts (stem, leaf and root) were separated and packed in a paper bag; they were then dried in an oven with forced air ventilation at a temperature of 65°C until a constant weight was obtained.

The material was then weighed and the phytomass of the leaves, stem, root and total was obtained.

The quality of the seedlings was estimated using the Dickson quality index (DQI) for seedlings, using the formula of Dickson et al. (1960), described by Equation 2.

$$IQD = \frac{FST}{(AP/DC) + (FSPA/FSR)} \tag{2}$$

In which:

IQD - Dickson's quality index;

AP - plant height (cm);

DC - stem diameter (mm);

FST - total plant dry phytomass (g);

FSPA - plant aerial part dry phytomass (g); and, FSR - plant root dry phytomass (g).

The data collected was subjected to analysis of variance using the F test at 0.05 and 0.01 probability levels and, when significant, linear and quadratic polynomial regression analysis was carried out using the SISVAR statistical software (Ferreira, 2014).

3. RESULTS AND DISCUSSION

The summary of the analysis of variance (Table 2) shows that the relative growth rates in height and diameter, the accumulation of dry phytomass of different parts of the plant and the quality of the Morada Nova soursop seedlings were not influenced ($p > 0.05$) by the interaction between salt levels and hydrogen peroxide (NS x H_2O_2). The salt levels in the irrigation water had a significant influence ($p < 0.01$) on all the variables studied, with the exception of the relative growth rate in plant height (TCRap). The concentrations of H_2O_2, on the *other hand, had a* significant effect on the relative growth rate in leaf area (TCRaf), leaf dry phytomass (FSF) and stem dry phytomass (FSC).

Table 2. Summary of the F test for relative growth rates in plant height (TCRap), stem diameter (TCRdc), and leaf area (TCRaf) from 85 to 145 days after sowing (DAS) and, dry phytomass of leaves (FSF), stem (FSC), root (FSR) and total (FST), and Dickson quality index (IQD) at 145 DAS of soursop seedlings cv. Morada Nova irrigated with saline water and exogenous application of hydrogen peroxide.

Treatments	F-test							
	TCRap	TCRdc	TCRaf	FSF	FSC	FSR	FST	IQD
Salt levels (NS)	ns	**	**	**	**	**	**	**
Linear regression	ns	**	ns	**	**	**	**	**
Quadratic regression	ns	ns	**	ns	ns	ns	ns	ns
Hydrogen peroxide (H_2O_2)			**	*	**			ns
Linear regression	ns	ns	**	**	**	**	**	ns
Quadratic regression	ns	ns	**	ns	ns	ns	ns	**
Interaction (NS x H2O2)	ns	ns	ns	ns	ns	ns	ns	ns
Blocks	ns	ns	ns	ns	ns	ns	ns	ns
CV (%)	12,06	15,78	15,89	23,02	21,52	28,67	22,09	24,52

ns, **, * respectively not significant, significant at $p < 0.01$ and $p < 0.05$

The regression equation (Figure 1) obtained for the relative growth rate of stem diameter shows a reduction of 7.57% per unit increase in irrigation water salinity; Thus, plants irrigated with 3.5 dS m water^{-1} had a 21.21% reduction in TCRdc compared to plants irrigated with 0.7 dS m water^{-1} , i.e. the diameter of plants irrigated with the lowest salinity level grew 0.003 mm mm^{-1} per day more than plants cultivated with the highest salinity level, between 85 and 145 DAS.

The salinity of the irrigation water negatively affected plant growth, possibly due to the osmotic and specific effects of the ions, slowing down cell expansion and division, promoting negative consequences on the photosynthetic rate, damaging the physiological and biochemical processes of the plants (Bezerra et al., 2018). Souza et al. (2017), studying guava, also observed a decrease in TCRdc under conditions of saline stress (ECa ranging from 0.3 to 3.5 dS m-1), with a reduction of 5.31% per unit increase in ECa.

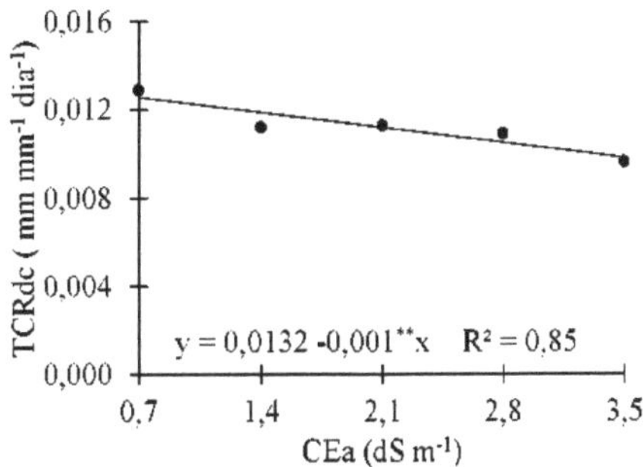

Figure 1 - Relative growth rate in stem diameter (TRCdc) of soursop cv. Morada Nova as a function of irrigation water salinity, from 85 to 145 days after sowing.

For the relative growth rate in leaf area (TCRaf), there is a decreasing linear trend in TCRaf with increasing salinity of the irrigation water and, according to the regression equation (Figure 2A), irrigation with a water electrical conductivity level of 0.7 dS m^{-1} promoted a relative growth rate in leaf area of 0.022 cm cm$^{2-2}$ day^{-1} ; However, the plants irrigated with water of 3.5 dS m^{-1} had a rate of 0.013 cm2 cm^{-2} day^{-1} , i.e. there was a reduction of 0.009 cm2 cm^{-2} day^{-1} (35.72%) in the plants irrigated with the highest saline level, compared to those grown with the lowest saline level.

Comparatively, the order of the degenerative effects of irrigation water salinity on the morphological behavior of soursop was greater in the TCRaf than in the TCRdc of the seedlings, which expresses the greater sensitivity of the leaves to salinity, showing that different organs of the plants can respond differently to the action of salts. In addition, this reduction may be related to the unbalanced absorption of nutrients by plants grown under conditions of saline stress, which causes damage to leaf tissues, leading to more pronounced inhibition of leaf elongation, decreasing the transpiration surface and resulting in a reduction in water absorption by the plants, since under such conditions it is interesting to see a reduction in transpiration and, consequently, a reduction in the loading of Na^{+} and Cl- into the xylem and water conservation in the plant tissues (Munns & Test, 2008).

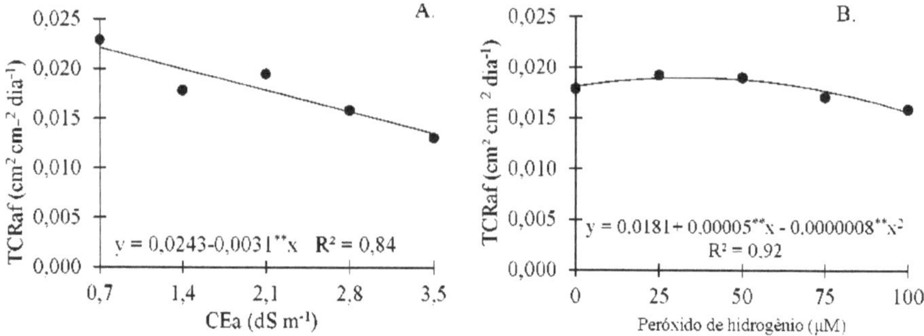

Figure 2: Relative growth rate of leaf area (RGFR) of soursop cv. Morada Nova as a function of irrigation water salinity - ECa (A) and hydrogen peroxide - HO_{22} (B) from 85 to 145 days after sowing.

The regression equation (Figure 2B) allows us to estimate the effect of hydrogen peroxide concentration on TCR_{af} between 85 and 145 DAS. Thus, a quadratic response was observed for TCR_{af} with a positive effect of H2O2 application up to a concentration of 31 µM of H2O2 (0.0189 cm^2 cm^{-2} day^{-1}); from this point on, there was a reduction in TCR_a f, reaching the lowest estimated value of 0.0151 cm2 cm^{-2} day^{-1} in the plants that received 100 µM of H2O2. Based on the results, it can be inferred that adequate applications of H2O2 can promote greater growth in plants under stress, as H2O2 can stimulate the accumulation of proteins and soluble carbohydrates, which will act as organic solutes, carrying out the osmotic adjustment of plants under conditions of saline stress, culminating in greater absorption of water and nutrients. In addition, it can minimize the effects of salinity on stomatal conductance, resulting in the proper physiological functioning of the plant (Carvalho et al., 2011).

The regression equations obtained for the dry phytomass of the stem (FSC) and leaves (FSF) of the soursop cv. Morada Nova (Figure 3A), there is a linear and decreasing effect, with a decrease of 14.04% and 14.24%, respectively, in SCF and SCF, per unit increase in ECa, equivalent to a reduction of 0.48 and 0.74 g in the plants subjected to irrigation with water of the highest salinity level (3.5 dS m-1), when compared to those under the lowest ECa level of 0.7 dS m-1. The reduction in dry phytomass is closely linked to the effects of the accumulation of soluble salts, which is a limiting factor in the development of most crops, and it can be deduced that this behavior can be understood as a possible plant adjustment mechanism to reduce the effects of salinity; in addition to changes in the ionic balance, water potential, mineral nutrition, stomatal closure, photosynthetic efficiency and carbon allocation, these factors cause morphological and/or physiological changes in plants, such as a reduction in biomass, when they are subjected to saline stress (Centeno et al, 2014). Sà et al. (2015), when studying the initial growth of pine trees under irrigation with saline water, also found a decrease in plant dry mass with increasing water salinity and, in turn, attributed this result to the osmotic and ionic effects caused by salinity.

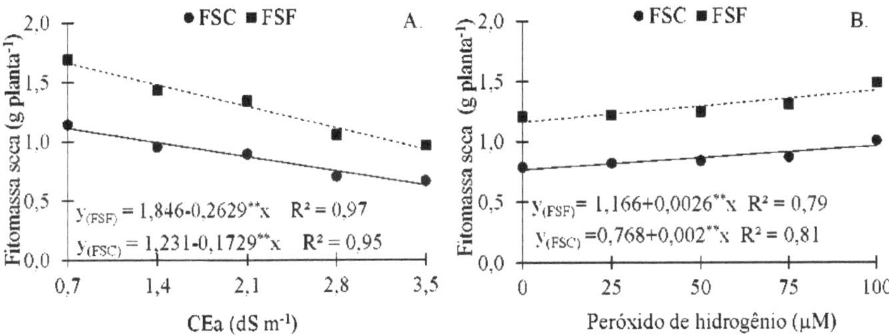

Figura 3. Leaf dry phytomass - FSF and stem dry phytomass - FSC of Morada Nova soursop, as a function of irrigation water salinity - ECa (A) and hydrogen peroxide concentrations - H2O2 (B), at 145 days after sowing.

With regard to the effects of hydrogen peroxide concentrations on the formation of mass in Morada Nova soursop (Figure 3B), the regression equations show a linear increase in SCF and SCF, with increases of 26.04% and 22.29%, respectively, in the plants that received the maximum concentration (100 µM) of H2O2 compared to the control treatment (0 µM). It can be seen from this that the application of H2O2, with the aim of it acting as a mitigant to salt stress, was satisfactory, because when the plant is pre-exposed to moderate stresses or to signaling metabolites such as H2O2, it can result in metabolic signaling in the cell (increases in metabolites and/or antioxidative enzymes) and, therefore, in better physiological and metabolic performance when the plant is exposed to more severe stress conditions (Forman et al., 2010), resulting in greater tolerance to salt stress, 2010), resulting in greater plant tolerance to stress.

The root dry mass (RDS) and total dry mass (TDS) of soursop seedlings cv. Morada Nova were negatively affected by the increase in salinity of the irrigation water and through the regression equations (Figures 4A and B), the data fitted the linear model, whose estimated decreases were 19.74 and 15.96%, respectively, per unit increase in the electrical conductivity of the irrigation water, i.e. there was a reduction of approximately 0.82 g in the FSR and 2.04 g in the FST of the soursop seedlings cv. Morada Nova seedlings irrigated with water with a higher ECa level (3.5 dS m^{-1}), compared to those that received an ECa of 0.7 dS m^{-1} . The reduction in phytomass formation as a result of the stress caused by water salinity may be a plant tolerance strategy aimed at reducing the absorption of toxic ions, enabling ion homeostasis in plant metabolism (Sà et al., 2013). In addition, plants under salt stress tend to seek osmotic adjustment; however, this activity requires a great deal of energy which will be used to accumulate sugars, organic acids and ions in the vacuole, energy which under normal conditions could be converted into phytomass production (Santos et al., 2012).

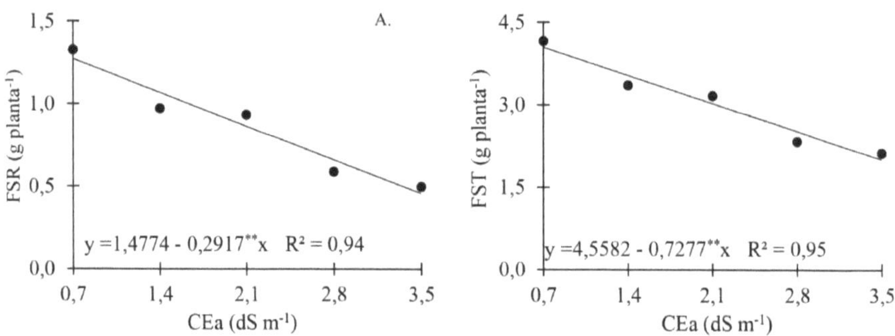

Figura 4. Root dry phytomass - FSR (A) and total dry phytomass - FST (B) of Morada Nova soursop as a function of irrigation water salinity at 145 days after

sowing.

The quality of Morada Nova soursop seedlings (IQD) was negatively affected by the increase in salinity of the irrigation water and the regression equation (Figure 5) shows a decreasing linear effect, with decreases of around 18.08% in IQD per unit increase in ECa at 145 DAS. When comparing the plants irrigated with water with the highest saline level (3.5 dS m^{-1}) in relation to those grown with an ECA of 0.7 dS m^{-1} , there was a 50.62% decrease in the IQD. According to Gomes (2001) and Oliveira et al. (2013), the QDI is an important morphological parameter used to express the quality and hardiness of seedlings by assessing their capacity for growth and survival. This was observed in this study where even the plants irrigated with the highest saline level (3.5 dS m^{-1}) had acceptable QDIs, bearing in mind that seedlings with QDIs above 0.2 are considered to be of good quality; Furthermore, the higher the IQD value, the better the quality of the seedling, as it expresses robustness and balance in the distribution of biomass.

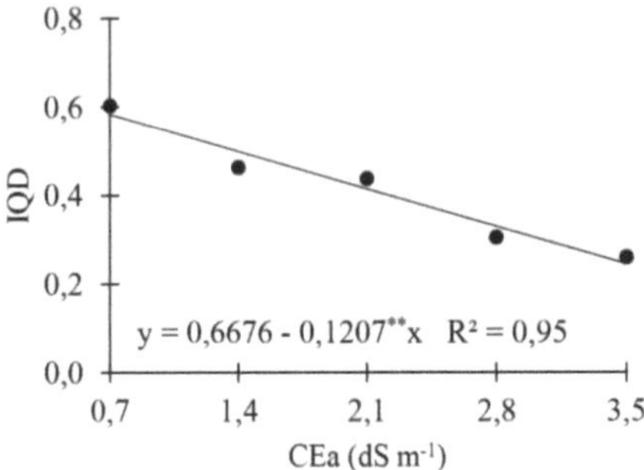

Figura 5. Dickson quality index - DQI of Morada Nova soursop as a function of irrigation water salinity at 145 days after sowing.

4. CONCLUSIONS

Immersion in water with an electrical conductivity above 0.7 dS m^{-1} negatively affects the growth of Morada Nova soursop, with root dry phytomass being the most sensitive variable.

Concentrations of 31 and 100 µM of hydrogen peroxide promote higher values for the relative growth rate in leaf area and leaf and stem dry matter of Morada Nova soursop seedlings, respectively.

Water with an electrical conductivity of 1.22 dS m^{-1} can be used to supply water to Morada Nova soursop seedlings, as it promotes an acceptable average reduction in growth of only 10%.

The quality of Morada Nova soursop seedlings was not compromised by using water with an electrical conductivity of 3.5 dS m^{-1} for irrigation, as the Dickson quality index was higher than 0.2.

5. BIBLIOGRAPHICAL REFERENCES

Ahmad, I.; Basra, S. M. A.; Afzal, I.; Farooq, M.; Wahid A. Stand establishment improvement in spring maize through exogenous application of ascorbic acid, salicylic acid and hydrogen peroxide. International Journal of Agriculture & Biology, v.15, p. 95-100, 2013.

Aragao, G. F.; Gomes-Filho, E.; Camelo, M. E.; Tarquinio, P. O. J. Effects of H O$_{22}$ on growth and solute accumulation in maize plants under salt stress. Revista Ciência Agrárias, v.42, p.373-381, 2011.

Benincasa, M. M. P. Plant growth analysis, basics. 2 ed. Jaboticabal: FUNEP, 2003. 41p.

Bezerra, I. L.; Nobre, R. G.; Gheyi, H. R.; Souza, L. P.; Pinheiro, F. W. A.; Lima, G. S. Morphophysiology of guava under saline water irrigation and nitrogen fertilization. Brazilian Journal of Agricultural and Environmental Engineering, v. 22, p. 32-37, 2018.

Carvalho, F. E. L.; Lobo, A. K. M.; Bonifacio, A.; Martins, M. O.; Neto, M. C. L.; Silveira, J. A. G. Acclimation to salt stress in rice plants induced by pretreatment with H2O2. Revista Brasileira de Engenharia Agricola e Ambiental, v.15, p.416-423, 2011.

Centeno, C. R. M.; Santos, J. B. dos; Xavier, D. A.; Azevedo, C. A. V. de; Gheyi, H. R. Production components of Embrapa 122-V2000 sunflower under water salinity and nitrogen fertilization. Revista Brasileira Engenharia Agricola Ambiental, v.18, (Suplemento), p. 39-45, 2014.

Dickson, A.; Leaf, A. L.; Hosner, J. F. Quality appraisal of white spruce and white pine seedling stock in nurseries. The Forest Chronicle, v. 36, n. 1, p. 10-13, 1960.

Donagema, G. K.; Campos, D. V. B. de; Calderano, S. B.; Teixeira, W. G.; Viana, J. H. M. (Org.). Manual of soil analysis methods. 2.ed. Rio de Janeiro: Embrapa Solos, 2011.

Ferreira, D. F. Sisvar: a computer statistical analysis system. Ciência e Agrotecnologia, v. 38, p. 109-112, 2014.

Freitas, A. L. G. E.; VILASBOAS, F. S.; PIRES, M. M.; SÂO JOSÉ, A. R. Characterization of graviola (*Annona muricata* L.) production and market in the state of Bahia. Informações Econômicas, v. 43, p.23-34, 2013.

Forman, H. J.; Maiorino, M.; Ursini, F. Signaling. functions of reactive oxygen species. Biochemistry, v.49, p.835-842, 2010.

Gomes, J. M. Morphological parameters in the evaluation of the quality of Eucalyptus grandis seedlings, produced in different sizes of tubing and dosages of N-P-K. (Doctoral Thesis). Federal University of Viçosa, Viçosa, p. 112, 2001

Holanda, J. S.; Amorim, J. R. A.; Ferreira Neto, M.; Holanda, A. C.; Sà, F. V. S. Water Quality for Irrigation. In: GHEYI, H. R.; DIAS, N.S.; LACERDA, C. F.; GOMES FILHO, E. (Org.). Qualidade da Agua para Irrigaçào. 2ed.Fortaleza CE: Instituto Nacional de Ciência e Tecnologia em Salinidade - INCTSal, 2016, v. 1, p. 35-50.

iseri, O. D.; Korpe D. A.; Sahin F. I.; Haberal M. Hydrogen peroxide pretreatment of roots enhanced oxidative stress response of tomato under cold stress. Acta Physiologiae Plantarum, v.35, p. 1905-1913, 2013.

Munns, R.; Tester, M. Mechanisms of salinity tolerance. Annual Review of Plant Biology, v.59, p.651-681, 2008.

Medeiros, J. F. de. Irrigation water quality and salinity evolution in properties assisted by GAT in the states of RN, PB and CE. (Master's dissertation). Federal University of Paraiba, Campina Grande. 2003, 173p.

Mendonça, V.; Ramos, J. D.; Araujo Neto, S. E.; Pio, R.; Contijo, T. C. A; Junqueira, K. P. Substrate and breaking of seed dormancy in the formation of soursop rootstock Cv. RBR. Revista Ceres, v. 49, p.657-668, 2002.

Neves, A. L. R.; Lacerda, C. F.; Guimaràes, F. V. A.; Hernandez, F. F. F.; Silva, F. B.; Prisco, J. T.; Gheyi, H. R. Biomass accumulation and nutrient extraction by string bean plants irrigated with saline water at different stages of development. Ciência Rural, v.39, p. 758-765, 2009.

Novais, R. F.; Neves J. C. L.; Barros N. F. Controlled environment trial. In: OLIVEIRA A. J. (ed). Research methods in soil fertility. Brasilia: Embrapa-SEA. p. 189-253. 1991.

Oliveira, F. T.; Hafle, O. M.; Mendonça, V.; Moreira, J. N.; Pereira Jûnior, E. B. Organic sources and container volumes on the initial growth of guava rootstocks. Revista Verde de Agroecologia e Desenvolvimento Sustentâvel, v. 7, p. 97-103, 2013.

Sâ, F. V. S. da; Brito, M. E. B.; Pereira, I. B.; Neto, P. A.; Andrade Silva, L. de; Costa, F. B. da. Salt balance and initial growth of pine seedlings (*Annona squamosa* L.) under substrates irrigated with saline water. Irriga, v.20, p. 544, 2015.

Sâ, F. V. S.; Brito, M. E. B.; Melo, A. S.; Antônio Neto, P.; Fernandes, P. D.; Ferreira, I. B. Production of papaya seedlings irrigated with saline water. Revista Brasileira Engenharia Agricola e Ambiental, v. 17, p. 1047-1054, 2013.

Sacramento, C. K.; Moura, J. I. L.; Coelho Junior, E. Graviola. In: SEREJO, J. A. dos SANTOS; DANTAS, J. L. L.; SAMPAIO, C. V.; COELHO, Y. da S. (eds.). Tropical fruit growing: regional and exotic species. Brasilia, DF: Embrapa Informaçâo Tecnológica, p. 95-132, 2009.

Santos, B. dos; Ferreira, P. A.; Oliveira, F. G. de; Batista, R. O.; Costa, A. C.; Cano,

M. A. O.; Production and physiological parameters of peanut as a function of salt stress. Idesia, v.30, p.69-74, 2012.

Souza, L. de P.; Sena, G. Sâ A. de; Nobre, R. G.; Barbosa, J. L.; Souza; C. M. A. de; Elias, J. J. Guava rootstock formation submitted to different water salinities and nitrogen fertilization. Revista Brasileira de Agricultura Irrigada, v.11, p. 1578 - 1587, 2017.

Printed by Books on Demand GmbH, Norderstedt / Germany